ASPECTS OF QUANTUM THEORY

An early portrait of P. A. M. Dirac

(Copyright Ramsey and Muspratt Ltd, Cambridge)

Aspects of
Quantum Theory

Edited by
Abdus Salam
and
E. P. Wigner

Cambridge
at the University Press

CAMBRIDGE UNIVERSITY PRESS
Cambridge, New York, Melbourne, Madrid, Cape Town, Singapore,
São Paulo, Delhi, Dubai, Tokyo

Cambridge University Press
The Edinburgh Building, Cambridge CB2 8RU, UK

Published in the United States of America by Cambridge University Press, New York

www.cambridge.org
Information on this title: www.cambridge.org/9780521131032

© Cambridge University Press 1972

First published 1972
This digitally printed version 2010

A catalogue record for this publication is available from the British Library

Library of Congress Catalogue Card Number: 72-75298

ISBN 978-0-521-08600-4 Hardback
ISBN 978-0-521-13103-2 Paperback

Dedicated to

P. A. M. DIRAC

*to commemorate his seventieth birthday and his
contributions to quantum mechanics*

Contents

Plates

Preface

On the 8 August 1972 Paul Adrien Maurice Dirac will be seventy. To celebrate this occasion, some of his pupils and admirers have prepared this volume of essays. Dirac is one of the chief creators of quantum mechanics. By concentrating on just those areas of quantum theory with which he is primarily associated, we have in fact been able to range over almost all its aspects.

Posterity will rate Dirac as one of the greatest physicists of all time. The present generation values him as one of its great teachers – teaching both through his lucid lectures as well as through his book *The Principles of Quantum Mechanics*. This exhibits a clarity and a spirit similar to those of the *Principia* written by a predecessor of his in the Lucasian Chair in Cambridge. On those privileged to know him, Dirac has left his mark, not only by his observations (which he makes rarely but which are always incisive), but even more by his human greatness. He is modest, affectionate, and sets the highest possible standards of personal and scientific integrity. He is a legend in his own lifetime and rightly so.

On behalf of all those who have contributed, we offer Dirac this volume as a token of our affection and gratitude.

ADDUS SALAM
EUGENE P. WIGNER

Contributors

PROFESSOR E. AMALDI, Istituto di Fisica 'Guglielmo Marconi', Rome.

PROFESSOR N. CABIBBO, Istituto di Fisica 'Guglielmo Marconi', Rome.

PROFESSOR FREEMAN J. DYSON, Institute for Advanced Study, Princeton, N.J.

DR R. J. EDEN, Cavendish Laboratory, University of Cambridge.

PROFESSOR W. HEISENBERG, Max-Planck-Institut für Physik and Astrophysik, Munich.

PROFESSOR J. M. JAUCH, Institut de Physique Theorique, Geneva.

PROFESSOR RES JOST, Eidgenöissische Technische Hochschule, Zurich.

PROFESSOR C. LANCZOS, Dublin Institute for Advanced Studies, Dublin.

DR JAGDISH MEHRA, Center for Particle Theory, University of Texas at Austin, Austin, Texas.

PROFESSOR A. PAIS, National Accelerator Laboratory, Batavia, Illinois.

PROFESSOR RUDOLPH PEIERLS, Department of Theoretical Physics, University of Oxford.

PROFESSOR J. C. POLKINGHORNE, Department of Applied Mathematics and Theoretical Physics, University of Cambridge.

PROFESSOR ABDUS SALAM, International Centre for Theoretical Physics, Trieste, and Imperial College of Science and Technology, London.

PROFESSOR L. SCHWARTZ, Ecole Polytechnique, Paris.

DR J. STRATHDEE, International Centre for Theoretical Physics, Trieste.

PROFESSOR J. H. VAN VLECK, Harvard University, Cambridge, Mass.

PROFESSOR A. S. WIGHTMAN, Department of Physics, Princeton University.

PROFESSOR EUGENE P. WIGNER, Department of Physics, Princeton University.

Bibliography of P. A. M. Dirac

[1] Dissociation under a temperature gradient, *Proc. Cambridge Phil. Soc.* **22**, 132–7 (1924).

[2] Note on the relativity dynamics of a particle, *Phil. Mag.* **47**, 1158–9 (1924).

[3] Note on the Doppler principle and Bohr's frequency condition, *Proc. Cambridge Phil. Soc.* **22**, 432–3 (1924).

[4] The conditions for statistical equilibrium between atoms, electrons and radiation, *Proc. Roy. Soc.* (*London*) A**106**, 581–96 (1924).

[5] The adiabatic invariants of the quantum integrals, *Proc. Roy. Soc.* (*London*) A**107**, 725–34 (1925).

[6] The effect of Compton scattering by free electrons in a stellar atmosphere, *Monthly Notices Roy. Astron. Soc.* (*London*) **85**, 825–32 (1925).

[7] The adiabatic hypothesis for magnetic fields, *Proc. Cambridge Phil. Soc.* **23**, 69–72 (1925).

[8] The fundamental equations of quantum mechanics, *Proc. Roy. Soc.* (*London*) A**109**, 642–53 (1925).

[9] Quantum mechanics and a preliminary investigation of the hydrogen atom, *Proc. Roy. Soc.* (*London*) A**110**, 561–79 (1926).

[10] The elimination of the nodes in quantum mechanics, *Proc. Roy. Soc.* (*London*) A**111**, 281–305 (1926).

[11] Relativity quantum mechanics with an application to Compton scattering, *Proc. Roy. Soc.* (*London*) A**111**, 405–23 (1926).

[12] Quantum Mechanics, Cambridge University dissertation, May 1926.

[13] On quantum algebra, *Proc. Cambridge Phil. Soc.* **23**, 412–18 (1926).

[14] On the theory of quantum mechanics, *Proc. Roy. Soc.* (*London*) A**112**, 661–77 (1926).

[15] The Compton effect in wave mechanics, *Proc. Cambridge Phil. Soc.* **23**, 500 7 (1026).

[16] The physical interpretation of the quantum dynamics, *Proc. Roy. Soc.* (*London*) A**113**, 621–41 (1927).

[17] The quantum theory of emission and absorption of radiation, *Proc. Roy. Soc.* (*London*) A**114**, 243–65 (1927).

[18] The quantum theory of dispersion, *Proc. Roy. Soc.* (*London*) A**114**, 710–28 (1927).

[19] Über die Quantenmechanik der Stossvorgänge, *Z. Physik* **44**, 585–95 (1927).

[20] The quantum theory of the electron, I, *Proc. Roy. Soc.* (*London*) A**117**, 610–24 (1928).

[21] The quantum theory of the electron, II, *Proc. Roy. Soc.* (*London*) A**118**, 351–61 (1928).

[22] Über the Quantentheorie des Elektrons, *Phys. Zeitschr.* **29**, 561–3 (1928). (Report on Dirac's lecture at the 'Leipziger Universitätswochen', 18–23 June 1928.)

[23] The basis of statistical quantum mechanics, *Proc. Cambridge Phil. Soc.* **25**, 62–6 (1929).

[24] Quantum mechanics of many electron systems, *Proc. Roy. Soc.* (*London*) A**123**, 714–33 (1929).

[25] A theory of electrons and protons, *Proc. Roy. Soc.* (*London*) A**126**, 360–5 (1930).

[26] On the annihilation of electrons and protons, *Proc. Cambridge Phil. Soc.* **26**, 361–75 (1930).

[27] Note on exchange phenomena in the Thomas atom, *Proc. Cambridge Phil. Soc.* **26**, 376–85 (1930).

[28] The proton, *Nature* **126**, 605 (1930).

[29] *The Principles of Quantum Mechanics* (Clarendon Press, Oxford: 1930).

[30] Note on the interpretation of the density matrix in the many electron problem, *Proc. Cambridge Phil. Soc.* **27**, 240–3 (1931).

[31] Quantized singularities in the electromagnetic field, *Proc. Roy. Soc. (London)* A**133**, 60–72 (1931).

[32] Photo-electric absorption in hydrogen-like atoms (with J. W. Harding), *Proc. Cambridge Phil. Soc.* **28**, 209–18 (1932).

[33] Relativistic quantum mechanics, *Proc. Roy. Soc. (London)* A**136**, 453–64 (1932).

[34] On quantum electrodynamics (with V. A. Fock and B. Podolsky), *Phys. Zeitschr. der Sowjetunion* **2**, 468–79 (1932).

[35] The Lagrangian in quantum mechanics, *Phys. Zeitschr. der Sowjetunion* **3**, 64–72 (1933).

[36] The reflection of electrons from standing light waves, with P. Kapitza, *Proc. Cambridge Phil. Soc.* **29**, 297–300 (1933).

[37] Homogeneous variables in classical dynamics, *Proc. Cambridge Phil. Soc.* **29**, 389–401 (1933).

[38] *Théorie du Positron, Septiéme Conseil de Physique Solvay* (*Structure et Propriétés des Noyaux Atomiques*), 22–29 October 1933 (Gauthier-Villars, Paris: 1934).

[39] Theory of electrons and positrons, *Nobel Lectures – Physics 1922–41*, pp. 320–5 (Amsterdam: 1965).

[40] Discussion of the infinite distribution of electrons in the theory of the positron, *Proc. Cambridge Phil. Soc.* **30**, 150–63 (1934).

[41] Does conservation of energy hold in atomic processes? *Nature* **137**, 298–9 (1936).

[42] Relativistic wave equations, *Proc. Roy. Soc. (London)* A**155**, 447–59 (1936).

[43] The cosmological constants, *Nature* **139**, 323 (1937).

[44] Physical science and philosophy, *Nature* **139**, 1001–2 (1937).

[45] Complex variables in quantum mechanics, *Proc. Roy. Soc. (London)* A**160**, 48–59 (1937).

[46] A new basis for cosmology, *Proc. Roy. Soc. (London)* A**165**, 199–208 (1938).

[47] Classical theory of radiating electrons, *Proc. Roy. Soc. (London)* A**167**, 148–69 (1938).

[48] The relation between mathematics and physics (James Scott Prize Lecture) *Proc. Roy. Soc. (Edinburgh)* **59**, 122–9 (1939).

[49] A new notation for quantum mechanics, *Proc. Cambridge Phil. Soc.* **35**, 416–18 (1939).

[50] La théorie de l'électron et du champ éléctromagnetique, *Ann. Inst. H. Poincaré* **9**, 13–49 (1939).

[51] Dr M. Mathisson (Obituary), *Nature* **146**, 613 (1940).

[52] The physical interpretation of quantum mechanics (Bakerian Lecture 1941), *Proc. Roy. Soc. (London)* A**180**, 1–40 (1942).

[53] On Lorentz invariance in the quantum theory (with R. Peierls and M. H. L. Pryce), *Proc. Cambridge Phil. Soc.* **38**, 193–200 (1942).

[54] Quantum electrodynamics, *Comm. Dublin Inst. Adv. Stud.* ser. A, no. 1 (1943).

[55] Unitary representations of the Lorentz group, *Proc. Roy. Soc. (London)* A183, 284–95 (1945).

[56] On the analogy between classical and quantum mechanics, *Rev. Mod. Phys.* 17, 195–9 (1945).

[57] Applications of quaternions to Lorentz transformations, *Proc. Roy. Irish Acad. (Dublin)* A50, 261–70 (1945).

[58] Developments in quantum electrodynamics, *Comm. Dublin Inst. Adv. Stud.* ser. A, no. 3 (1946).

[59] On the theory of point electrons, *Phil. Mag.* 39, 31–4 (1948).

[60] The difficulties in quantum electrodynamics, *Rep. Int. Conf. on Fundamental Particles and Low Temperatures, July 1946*, vol. 1, pp. 10–14 (Physical Society, London: 1948).

[61] Quantum theory of localizable dynamic systems, *Phys. Rev.* 73, 1092–103 (1948).

[62] The theory of magnetic poles, *Phys. Rev.* 74, 817–30 (1948).

[63] Forms of relativistic dynamics, *Rev. Mod. Phys.* 21, 392–9 (1949).

[64] La seconde quantification, *Ann. Inst. H. Poincaré* 11, no. 1, 15–47 (1949).

[65] A new meaning for gauge transformations in electrodynamics, *Nuovo Cimento* (9) 7, 925–38 (1950).

[66] Generalized Hamiltonian dynamics, *Can. J. Math.* 2, 129–48 (1950).

[67] The Hamiltonian form of field dynamics, *Can. J. Math.* 3, 1–23 (1951).

[68] The relation of classical to quantum mechanics, *Proc. Second Canadian Math. Congress* (Vancouver 1949), pp. 10–31 (University of Toronto Press, Toronto: 1951).

[69] Is there an aether? *Nature* 168, 906–7 (1951).

[70] A new classical theory of electrons, I, *Proc. Roy. Soc. (London)* A209, 291–6 (1951).

[71] Is there an aether? *Nature* 169, 146, 702 (1952).

[72] A new classical theory of electrons, II, *Proc. Roy. Soc. (London)* A212, 330–9 (1952).

[73] Les transformations de jauge en électrodynamique, *Ann. Inst. H. Poincaré* 13, no. 1, 1–42 (1952).

[74] The Lorentz transformation and absolute time, *Proc. Lorentz–Kamerlingh Onnes Memorial Conference, Leiden 22–26 June 1953* (Amsterdam: 1953), reprinted in *Physica* 19, 888–96 (1953).

[75] A new classical theory of electrons, III, *Proc. Roy. Soc. (London)* A223, 438–45 (1954).

[76] Quantum mechanics and the aether, *Scientific Monthly* 78, 142–6 (1954).

[77] The stress tensor in field dynamics, *Nuovo Cimento* (10) 1, 16–36 (1955).

[78] Gauge-invariant formulation of quantum electrodynamics, *Can. J. Phys.* 33, 650–60 (1955).

[79] Note on the use of non-orthogonal wave functions in perturbation calculations, *Can. J. Phys.* 33, 709–12 (1955).

[80] The vacuum in quantum electrodynamics, *Suppl. Nuovo Cimento* (10) 6, 322–39 (1957).

[81] Generalized Hamilton dynamics, *Proc. Roy. Soc. (London)* A246, 326–32 (1958).

[82] The theory of gravitation in Hamiltonian form, *Proc. Roy. Soc. (London)* A246, 333–43 (1958).

[83] The electron wave equation in Riemannian space, in *Max-Planck-Festschrift 1958*, ed. B. Kockel, W. Macke, and A. Papapetrou, pp. 339–44 (Verlag der Wiss., Berlin: 1958).

[84] Fixation of coordinates in the Hamiltonian theory of gravitation, *Phys. Rev.* **114**, 924–30 (1959).
[85] Energy of the gravitational field, *Phys. Rev. Letters* **2**, 368–71 (1959).
[86] Gravitationswellen, *Phys. Blätter* **16**, 364–6 (1960).
[87] A reformulation of the Born-Infeld electrodynamics, *Proc. Roy. Soc. (London)* A**257**, 32–43 (1960).
[88] Prof. Erwin Schrödinger, For. Mem. R.S. (Obituary), *Nature* **189**, 355–6 (1961).
[89] The energy of the gravitational field, in *Les Théories Relativistes de la Gravitation* (Coll. Int. du CNRS at Royaumont 1959), ed. the CNRS (Paris: 1962).
[90] Interacting gravitational and spinor fields, in *Recent Developments in General Relativity*, pp. 191–200 (PWN-Polish Scientific Publications, Warsaw: 1962).
[91] An extensible model of the electron, *Proc. Roy. Soc. (London)* A**268**, 57–67 (1962).
[92] Reply to a letter of R. H. Dicke on 'Dirac's cosmology and Mach's principle', *Nature* **192**, 441 (1961).
[93] Particles of finite size in the gravitation field, *Proc. Roy. Soc. (London)* A**270**, 354–6 (1962).
[94] The conditions for a quantum field theory to be relativistic, *Rev. Mod. Phys.* **34**, 592–6 (1962).
[95] The evolution of the physicist's picture of nature, *Scientific American* **208**, no. 5, 45–53 (1963).
[96] A remarkable representation of the 3 + 2 de Sitter group, *J. Math. Phys.* **4**, 901–9 (1963).
[97] Foundations of quantum mechanics, *Nature* **203**, 115–16 (1964).
[98] Hamiltonian methods and quantum mechanics, *Proc. Roy. Irish Acad.* A**63**, 49–59 (1964).
[99] Equivalence of the Schrödinger and Heisenberg pictures (with H. S. Perlman) *Nature* **204**, 771–2 (1964).
[100] *Lectures on Quantum Mechanics* (Academic Press, New York: 1964).
[101] Quantum electrodynamics without dead wood, *Phys. Rev.* **139**B, 684–90 (1965).
[102] *Lectures on Quantum Field Theory* (Academic Press, New York: 1966).
[103] The versatility of Niels Bohr, pp. 306–9, in *Niels Bohr, His Life and Work* (North-Holland Publishing Co., Amsterdam: 1967).
[104] Methods in Theoretical Physics, Second Evening Lecture in the Series 'From a Life of Physics', at the International Symposium on Contemporary Physics, Trieste 1968, Special Suppl. of IAEA Bulletin, IAEA, Vienna 1969.
[105] Can equations of motion be used? *Coral Gables Conference on Fundamental Interactions at High Energy, Coral Gables, 22–24 Jan. 1969*, pp. 1–18 (Gordon and Breach, New York: 1969).
[106] Hopes and fears, *Eureka*, no. 32, October 1969, pp. 2–4.
[107] Can equations of motion be used in high energy physics?, *Physics Today* **23**, no. 4, 29–31 (1970).
[108] A positive energy relativistic wave equation, *Proc. Roy. Soc. (London)* A**322**, 435–45 (1971).
[109] *The Development of Quantum Theory* (Gordon and Breach, New York: 1971).

(Compiled by JAGDISH MEHRA)

I

Dirac in Cambridge

R. J. Eden and J. C. Polkinghorne

Paul Dirac held the Lucasian Chair of Mathematics at Cambridge from his appointment in 1932 at the age of 30 until his retirement in 1969. This professorship is the oldest in the Faculty of Mathematics – a faculty which includes many people working in theoretical physics as well as people working in pure mathematics – and among his predecessors was Isaac Newton.

He came to Cambridge first in 1923 as a graduate student after studying electrical engineering at Bristol, and became a Fellow of St John's College in 1927. His reputation was soon made by his paper in 1925 on the fundamental laws of quantum mechanics which gave the world a deep insight into the meaning and implications of Heisenberg's work on matrix mechanics. Heisenberg recalled later how he had visited Cambridge in 1925 to lecture on his new matrix mechanics and how much he was impressed by the penetrating questions asked by a young research student. Heisenberg added that he was even more impressed two months later when the same student sent him a thirty-page manuscript setting out the fundamental laws of quantum mechanics!† On another occasion Heisenberg remarked that at first he had looked on the lack of commutation of coordinates and momenta as a defect in the theory until he saw from Dirac's presentation that it was one of the central features of quantum mechanics. It is characteristic of Dirac that he once introduced a lecture by Heisenberg by expressing great admiration for the speaker which was due, he said, to the fact that they had both worked on the same problem (of inventing the new quantum

† Dirac's recollection of this historic occasion is that Heisenberg's Cambridge lecture was mostly on the Zeeman effect, in which he (Dirac) was not much interested, and he did not notice the part on matrix mechanics which probably came at the end. Dirac did pay attention however a few weeks later when he received via R. H. Fowler a proof-copy of Heisenberg's paper. It was then that he realized the significance of this work and its relation to his own approach which very quickly lead to his paper on the fundamental laws of quantum mechanics.

B

theory) and Heisenberg had succeeded in being the first to solve it, whereas he had failed!

In Cambridge in the 1920s research students worked in their own colleges, in comparative isolation. There were no regular theoretical seminars and it was unusual to know students in other colleges even if they were working in the same field. Despite his also being a Johnian, Mott recalls how he spent most of the year 1927 reading Dirac's papers on radiation theory but does not believe that they had any discussions on physics at that time. It was apparently not the fashion in those days to ask someone to help one to do theoretical physics research.

Concerning those early days Lady Jeffreys writes:

My memories of that time are very clear as I became a research student under the supervision of R. H. Fowler in 1925. At that time Fowler was the central figure in Theoretical Physics in Cambridge. Quantum Mechanics was just getting under way. My notebook for 1926 contains:

Lent Term   Mr. Fowler   Quantum Theory of Spectra.
Easter Term   Mr. Fowler   Quantum Theory.
Easter Term   Mr. Dirac   Quantum *Mechanics*.

These lectures were all attended by the same small bunch of people. At this distance of time it is difficult to be sure but I think that it included D. R. Hartree, J. M. Whittaker, A. H. Wilson, J. A. Gaunt, N. F. Mott and J. R. Oppenheimer. The Easter term lectures were complementary to each other. Fowler had a remarkable gift for rapidly digesting new work of others and imparting his own enthusiasm for it. Dirac gave us what he himself had recently done, some of it already published, some, I think, not. We did not, it is true, form a very sociable group, but for anyone who was there it is impossible to forget the sense of excitement at the new work. I stood in some awe of Dirac, but if I did pluck up courage to ask him a question I always got a direct and helpful answer, with no beating about the bush if I was getting things wrong.

Whatever their deficiencies at that time as centres of research, the colleges at Cambridge have always succeeded in providing within the large university smaller communities in which social life can flourish. The bachelor tables of the 1920s were more hearty and bibulous than they are today and it is said that Dirac once showed his disapproval of that style of life by filling all the glasses at dinner with water which consequently had to be consumed before beer could be ordered.

He has always been a great walker and used to include some mountain climbing with his walking. Tamm visited Cambridge from the Soviet Union about 1931 and invited Dirac to go climbing with him on the Soviet–Chinese border. Mott recalls how Dirac and Tamm trained for this by climbing trees on the Gog-Magog hills near Cambridge, Dirac always wearing a formal dark suit. By this time Dirac had already invented the relativistic equation for the electron, which appears at first sight to indicate that a free electron has only two velocities namely $\pm c$, an appearance which Dirac explained at a deeper level which leads to the average velocity that arises in an actual observation. It is said that like his electron, Dirac's car in the early 1930s had only two velocities, one being full speed, the other being zero!

The extension into everyday life by Dirac of his logical approach to the problems of theoretical physics has yielded many stories. One of the earliest of these is recounted by Hulme who was Dirac's first research student. They used to go walking together in the Cambridge-shire countryside, apparently with little conversation and especially little on physics. Hulme remarks that they both used walking sticks and on one occasion he jokingly asked whether Dirac had ever tried walking with two sticks for symmetry. To his surprise, Dirac replied that he had tried once and it didn't work. He also remembers an occasion when he encountered Dirac going to London by train to the motor show. They met again on the return train to Cambridge, and Hulme was sucking some throat lozenges from a glass bottle that rattled a little in his pocket. Dirac asked if he had developed a sore throat during the day as he had not heard the lozenges rattling on the way to London. Hulme explained that the bottle had been full then so that they had not rattled, to which Dirac replied that he supposed when the bottle was half empty they would rattle most. Peierls has remarked how interesting it would have been if the story about the lozenges had preceded the hole theory of positrons; alas it is not so, the lozenges story is dated 1934 ± 2 so it was presumably derived *from* the hole theory (1931).

The penetrating directness of Dirac's papers and the relatively small number of references has sometimes been thought to suggest that he did not have wide interests in physics. Peierls remarks that during his time in Cambridge following 1933, Dirac appeared to him to be a person who could become interested in almost anything and Peierls illustrates this point by the following account of a little-known aspect of Dirac's research:

In about 1934 Dirac invented a method of isotope separation. The idea was to make a jet of gas turn a corner, past a sharp edge, so that the centrifugal force would cause separation of the components. He not only conceived the idea, but decided to verify it experimentally. Kapitza allowed him the use of a compressor in the Mond Laboratory, and the device was tried initially on a mixture, not of isotopes, but of air with a heavy organic compound. When I saw the experiment there had been as yet no evidence of a difference in composition between the two output tubes, but by feeling the tubes one could easily check that one was hot and the other cold, showing that something non-trivial was happening in the junction. When Kapitza had to stay in Moscow, and his equipment was sent on,

Dirac's experiment was interrupted. During the war, however, a group in Oxford studied the feasibility of the method for separating uranium hexafluoride. They found it worked perfectly well, but less efficiently than gaseous diffusion, and it was therefore not pursued. When this work was started, Dirac was invited to Oxford to discuss the method. I was present at the meeting in the Clarendon Laboratory (there may have been more than one meeting) and I remember that the experimentalists expected a highbrow and abstract mathematician who would know the kinetic theory of the effect, but would not know one end of an apparatus from another. They were most impressed by Dirac's eminently practical and helpful remarks.

Peierls continues,

Dirac's experiment also provided the occasion for one of my favourite episodes, though I do not know whether this is more a story about Dirac or about Wigner. Wigner and I happened to visit Cambridge on the same day, and we both called, separately, in the Mond Laboratory and saw Dirac with his apparatus. Later Wigner complained that Dirac was so secretive about his idea; he had refused to explain it. I was surprised by this, since my impression was different. I had actually guessed what the principle might be, but Dirac had shown no hesitation in confirming that my guess was right. A little more probing brought out that the relevant conversation consisted on one exchange. Wigner had said 'It must have been very difficult to make the little brass piece?' (A brass T piece, with the gas mixture entering at the stem and the fractions coming out of the arms, was clearly the heart of the device). Dirac's answer was 'No, that was fairly easy'. He had given a straight answer to a direct question. Wigner on the other hand, had asked for information and thought he had been refused. I bet Wigner that, by asking directly for the principle, he would have got the explanation. The bet could not be settled, because when we asked Dirac whether he would have responded to a direct question he said, of course, 'I do not know' – his frequent answer to a hypothetical question.

During the early and mid-1930s Dirac was 'adopted' first by the Mott family and then by the Peierls family during their successive periods of residence in Cambridge. In 1937 Dirac married Margit Wigner (sister of Eugene Wigner) and from that date the social life of a chosen few was enhanced by being 'adopted' by the Dirac family. It would be difficult to attempt to complete the list of those by whom the friendship of the Dirac family was much appreciated. Kemmer recalls, how the problems of secrecy in wartime work inhibited his scientific discussions with Dirac but he describes the friendly social contacts with the Dirac family during this time. Dirac was interested in gardening and Kemmer (like others after him) was often greeted by Mrs Dirac with 'Oh you want to see Paul? He is up a tree in the garden.'

At any given time, Dirac would not normally have more than one or two research students under his supervision. His reason was not at all related to the trouble involved, but was because his own interests were in fundamental problems and he did not think that these were suitable for many Ph.D. students. One of us (RJE) had the privilege of being in this small band and can testify that as a supervisor Dirac was most helpful, readily available to listen to any reports on progress. A seminar by one of his students would tend to take on the character of a public supervision (or inquisition) since he asked so many questions. He could be a formidable member of the audience if he chose. On one occasion the lecturer (a post-doctoral fellow) began by listing his five basic assumptions; when he had finished writing the fifth on the blackboard Dirac said 'Your assumption number two contradicts your assumption number five.' The lecturer paused for thought and then agreed. It says much for the resourcefulness of this particular speaker that he went on to say 'Since I cannot talk on that subject, I will give a seminar on a different topic', which he then proceeded to do.

Dirac's greatest influence on students in Cambridge generally was through his course of lectures on quantum theory. For many years it was the first course in quantum theory that Cambridge students could attend and many of us have had the privilege of learning our quantum mechanics 'straight from the horse's mouth', as the saying goes. Not all the audience were novices, however, for frequently visitors of some standing would rightly judge it something not to be missed while they were in Cambridge. The material and its treatment can readily be gauged from reading his celebrated book *The Principles of Quantum Mechanics*, which the course closely followed. However, there was more to the lectures than the printed page can convey. The delivery was always exceptionally clear and one was carried along in the unfolding of an argument which seemed as majestic and inevitable as the development of a Bach fugue. Gestures were kept to a minimum, though there was a celebrated passage near the beginning where he broke a piece of chalk in half and moving one of the bits about the lecture desk said that in quantum theory we must consider states which are a linear super-position of all these different possible locations. There was absolutely no attempt to underline what had been his own contributions, though at times one felt one got a hint of his feelings about what he had done. If there was substance in these perceptions they seemed to reveal that the invention of bra and ket vectors (and the naming of them) had given him as much pleasure as anything.

Dirac's tenure of the Lucasian Chair of Mathematics has given many generations of Cambridge men the honour and pleasure of knowing him as teacher, colleague and friend and they would wish to send him their warmest good wishes on his seventieth birthday.

Dirac at a recent Solvay Conference

2

Travels with Dirac in the Rockies

J. H. Van Vleck

It was in the spring of 1929 that Dirac made his first visit to the United
States, in response to an invitation from the University of Wisconsin to
spend a term as visiting professor there. I have always been proud of the
fact that my *alma mater* was the first institution to bring him to America
and that I participated in its decision to do so. Before proceeding to my
reminiscences of his early visits to my country, I will digress to tell how
even before then my research was influenced by his publications.

INFLUENCE OF DIRAC'S PAPERS ON MY EARLY WORK

During the first two years or so of quantum mechanics I was still at the
University of Minnesota, and my 'bibles' for learning the strange new
mechanics in early 1926 were the long papers by Born, Heisenberg and
Jordan in the *Zeitschrift für Physik*,[1] and Dirac's articles on 'The
fundamental equations of quantum mechanics', 'Quantum mechanics
and a preliminary investigation of the hydrogen atom', and 'The
elimination of the nodes in quantum mechanics', all in the *Proceedings
of the Royal Society*.[2] In another 'reminiscing' paper[3] I have described
how Dirac's work enabled me to compute the mean values of $1/r^2$ and
$1/r^3$ for a hydrogen atom (needed to calculate spin-relativistic fine
structure) and $1/r^4$ (needed to calculate quantum defects due to polarization in nearly hydrogenic atoms) and how when I reached Copenhagen
in 1926 with manuscripts all written, I found that these mean values
had just been published by Heisenberg and Jordan for $1/r^2$, $1/r^3$ and
were in course of publication for $1/r^4$ by Waller.

Hill and I[4] found formulas in Dirac's early publications which I have
cited also useful in connection with calculating matrix elements for fine
structures involving spin coupling in molecular spectra, a field that
would at first sight seem rather unrelated. I remember remarking to
Hill 'You can find almost anything in Dirac's papers if you read them
long enough.'

7

When Dirac's paper[5] on 'The quantum theory of the electron' reached me in Minneapolis in early 1928 it properly seemed to me sensational. The long article[6] on 'Magnetic susceptibilities in the new quantum mechanics' which I published in the *Physical Review* has for its second paragraph the following:

Note added in proof. A remarkable paper by Dirac (*Proc. Roy. Soc.*, Feb. 1928) has just appeared in which he shows that the requirement that the Schroedinger wave equation have the invariance demanded by relativity is adequate to give the terms ordinarily ascribed to internal spins of the electron. Thus our treatment of the electron as a spherical top to derive the Hamiltonian function inclusive of spin terms in a magnetic field suddenly loses much of its interest. However, it must at the same time be emphasized that all the essential results of the present paper are unaltered; the only difference is that our work prior to about p. 594 becomes rather antiquated, as Dirac's postulates give a Hamiltonian function such as (5) directly and elegantly. From there on everything goes as before.

In the 'antiquated pages prior to p. 594' I tried to describe the Uhlenbeck–Goudsmit spinning electron by means of a spherical top with Eulerian angles more or less as had Darwin[7] in an earlier paper written in 1927. In these pages there was a lot of nonsense relating to the electrical radius of the electron, in an attempt to show that there was no appreciable diamagnetic term due to spin and to make spin look as physical as possible in terms of classical concepts. My introductory model with Eulerian angles not merely wasted space in the *Physical Review*, but also appears to have diverted readers from the essential quantum-mechanical part of the calculation. At any rate, Sommerfeld[8] in his report to the 1930 Solvay congress appears to have overlooked the implications of my remarks added in proof, and gave the impression that my calculations, though giving the right answer, needed to be redone because of my use of Eulerian angles.

In a footnote of this same paper[6] I commented 'We use the matrix rather than wave formulation of the new quantum mechanics. The same results are, of course, obtained with either formulations in virtue of their general mathematical identity, and the popularity of susceptibility calculations by means of the wave equation seems rather surprising inasmuch as the matrix method has usually yielded the susceptibility formulas first and most directly.' As of today, this statement may appear a truism, but in the early days of quantum mechanics, things were so new that it was not easy for theoretical physicists to become thoroughly indoctrinated with both the matrix and differential equation approaches, even though they are both different aspects of general transformation theory. Unlike most theorists of the time, I belonged to a minority group preferring the former, but many papers of that era (e.g. part of Sommerfeld's report[8]) were written which simply redid with expansion of wave functions calculations previously made with matrix algebra. For instance, a 1928 paper of Darwin's[9] began as follows:

In a recent paper Dirac has brilliantly removed the defects before existing in the mechanics of the electron, and has shown how the phenomena usually called the 'spinning electron' fit into place in the complete theory. He applies to the problem the method of q-numbers and, using non-commutative algebra, exhibits the properties of a free electron, and of an electron in a central field of electric force. In a second paper he also discusses the rules of combination and the Zeeman effect. There are probably readers who will share the present writer's feeling that the methods of non-commutative algebra are harder to follow, and certainly much more difficult to invent, than are operations of types long familiar to analysis. Wherever it is possible to do so, it is surely better to present the theory in a mathematical form that dates from the time of Laplace and Legendre, if only because the details of the calculus have been so much more thoroughly explored.

However, Darwin's paper[9] was more than a mere transcription from matrix to wave-mechanical language; it contained the first proof that to all powers of $1/c^2$, Dirac's equations gave the same formula for the energy levels of hydrogen as that obtained by Sommerfeld in the old quantum theory with relativity but without spin. This is perhaps the most remarkable numerical coincidence in the history of physics; the physical interpretation and assignment of quantum numbers is, of course, completely different.

DIRAC IN MADISON, WISCONSIN 1929

When Dirac had been in Madison a few days I asked him what had surprised him most about the United States and he told me it was that there were so many wooden houses. I wonder whether this was responsible for his later wanting to travel extensively in the U.S.S.R., where there are also many wooden houses and where he once took the Trans-Siberian railway clear across that country.

Dirac gave a well-organized course of lectures, almost a formal course, on mainly the transformation theory of quantum mechanics which he had evolved about a year and a half earlier. It was far more abstract than the usual American courses in physics of that era. I remember my father, a Professor of Mathematics at the University of Wisconsin then in his final year before retirement, attending some of Dirac's lectures and remarking that their abstractness reminded him of the 'general analysis' which was developed by Hastings Moore, a mathematician at the University of Chicago and which was a very abstract theory of linear spaces.

Dirac either lectured on, or told me about his now celebrated vector model[10] for handling permutation degeneracy. In particular he indicated how it could be used to obtain Heisenberg's formulas for ferromagnetism, something not mentioned in his publications. I was greatly influenced by Dirac's procedure, and subsequently capitalized on it heavily, applying it not only to magnetism,[11] but also to complex spectra[12] and chemical bonding.[13] In consequence references are sometimes improperly made in the literature to the 'Dirac–Van Vleck vector model'. All I did was apply it. Dirac had the original and essential idea.

After Dirac had been in Madison only a short time, it became apparent that he was fond of walking, especially amid nice scenery. We used to take walks in the fields overlooking Lake Mendota, and elsewhere. Another of his attributes was a very fancy watch, which told the day of the week and month, and perhaps even the phase of moon (I can't remember). Its fame had spread to the University of Iowa, in advance of his giving a lecture there. One student or faculty member at Iowa City, hoping to see Dirac's watch, said 'Dr Dirac could you please tell me the time,' whereupon Dirac looked at the clock on one of the University buildings and gave him the time. One could claim that this reply was an illustration of Dirac's facility in answering a question in the most direct possible fashion. Professor Eldridge of the Iowa physics department, hearing of his colleague's failure, used the method of direct approach, and simply said to Dirac 'Show me that watch.'

One of the mysteries connected with Dirac's visit, almost as great as how he discovered his four component equations, was what his initials 'P.A.M.' stood for. He had signed his papers and correspondence only in this terse fashion. He was reluctant to take away the mystery and let people know what they stood for. Finally, perhaps near the end of Dirac's visit, there was a dinner in his honor at the University Club, for which Professor Ingersoll had a brilliant idea. Each place card contained not merely the name of who was to sit there, but also the statement that the dinner was in honor of . . . Dirac, and on each card was inserted a different guess as to what the 'P.A.M.' stood for. For instance one place card might read 'Professor Mendenhall, dinner in honor of Peter Alfred Martin Dirac.' After studying all the cards Dirac said that by proper combinations of names entered on certain of the cards, which numbered ten or so, one could obtain his real name (but as I remember it, one had to modify 'Morris' slightly to 'Maurice'). By considerable questioning as to which entries were right, we were finally able to learn that 'Paul Adrien Maurice' was the correct answer.

DIRAC AND HEISENBERG AROUND THE WORLD, 1929

Dirac decided that when he got as far west as Wisconsin he might as well go around the world, and after leaving Madison he made at least the first part of this trip along with Heisenberg. The latter was lecturing at the University of Chicago in the spring in 1929. While Dirac was in Madison Heisenberg came to Madison for a brief trip to give a colloquium. Whether they arranged their voyage then, or before Dirac left England, I don't know. There is one anecdote connected with their stay in Hawaii that I can't vouch for first hand, but I've heard it so often it must have some factual foundation. They decided to visit the university to pass some of the time when their boat laid over for a while in Honolulu. They managed to meet the chairman of the physics department and told him their names. This man looked at the two youthful visitors, and told them that if they would like to attend some of the physics lectures at the university while their boat was in Honolulu, they would be welcome to do so!

The only other piece of information I have about Dirac's trip around the world is that he stayed in the Kanaya Hotel in Nikko, Japan, in early September. This I know as a certainty for the following reason. After my father's retirement, my parents also travelled around the world. I remember father telling me later of Dirac's courtesy in crossing a very muddy street to shake hands with father when they were surprised to see each other in Nikko. In 1953, when I attended an international physics conference, I told the proprietor of the hotel in Nikko of the incident. He proudly produced the hotel register for 1929, and there we found the signatures of my father, of Dirac, and I believe, also of Heisenberg.

GLACIER NATIONAL PARK, MONTANA, 1931

No doubt when Dirac was in Madison in 1929 he heard me give glowing descriptions of the trails in Glacier National Park. At any rate we arranged to take a walking trip there in 1931. By that time other academic institutions (Princeton and Toronto?) were following in Wisconsin's footsteps and were inviting him to make extended visits to the

U.S.A., which he has done quite often ever since. We were favored by magnificent weather and we walked about seventy-five miles all told, taking the so-called 'North-Circle trip' via Granite Park and Flattop Mt. to Waterton Lake, and back via Indian Pass and Crossley Lake to the Many Glacier Hotel. During most of the time we stayed in permanent camps where we shared a tent together. He was an ideal tent-mate. Some physicists I have known (e.g. Fermi) liked to retire and arise early, whereas others (e.g. Pauli) preferred late hours at both ends. Dirac liked to retire early and arise late. He certainly never disturbed me in any way, and I only hope the reverse was also true. In the hotels he found the ice water very cold and the soup very hot by British standards, a situation he solved with his characteristic directness by transferring an ice cube from the water to the soup. I had to leave Glacier Park before Dirac did, and as I remember it he continued on to visit the Yellowstone Park.

COLORADO, 1934

I was immensely flattered when I received a letter from Dirac early in 1934, proposing that we take another walking trip in the Rockies. This time I suggested the San Juan Mountains region of Colorado, as this seemed to me more colorful than the Estes Park area, and it also had the advantage that it would give him a chance to visit Mesa Verde Park, which I have always considered one of America's wonders.

Dirac and I met at Kansas City. I had arranged for special permission from the Santa Fe Railroad for us to ride the 'California Limited' from there to La Junta, as in the pre-airplane age this train would not accept passengers who wanted to ride a distance of only 540 miles! (Otherwise to make our connection at La Junta we would have had to travel in a coach without a diner, on the rear end of a mail train.) The next day we rode the narrow gauge railroad over Marshall pass which later followed the Canyon of the Gunnison until it turned inland. Dirac decided he would like to go by foot for a distance further down the Gunnison than the railroad went. We thought we would have to walk back from Cimarron, the nearest station, some three miles inland, to get to the river, but when the conductor heard of our projected trip, he arranged for the engineer to stop the train where it left the river, and he said that the engineer would remember to stop to pick us up when the train returned that afternoon. What's more, he did! We spent the night at Gunnison, but the next day on our travels on the bus to a place called Lake City we did not fare as well. The advertized bus turned out to be really a mail truck with only two seats besides the driver. He assigned these to two individuals who were authentic westerners. Dirac and I were shut in with the mail sacks. They were uncomfortable to sit on as makeshift chairs for a run of some forty-five miles over a rough road. Furthermore, being in the dark, we were unable to enjoy any of the scenery.

At Lake City we each had a separate room in the hotel for 75 cents! When it was learned in this little hamlet (which was once a prosperous city during the silver mining boom) that there was *an Englishman* in town, it was real news, so a woman reporter from the local paper, which probably appeared once a week, came around to interview Dirac and get his impressions of Colorado. It might be a good project for some historian of science to ferret out the appropriate publication, if it can be done, and see what it said[14]. He had received the Nobel prize the year previous but I'm sure the reporter was not aware of this fact, nor would this have meant a thing to her if she were. Conversely, Dirac and I were not aware of perhaps Lake City's greatest claim to fame. Reputedly, during a starvation winter in the nineteenth century one inhabitant there kept himself alive by practising cannibalism – at any rate the 'Alfred Packard Memorial Grill' at the University of Colorado at Boulder was recently so named in his honor. Instead one reason we visited Lake City was to go to the top of Uncompaghre Peak, altitude 14 306 ft, which was accessible by horse almost the entire way. The other reason we went there was because I had read somewhere that a nice trip was to take an automobile up a rough road to a top of a pass, and then walk down to Ouray on an ill-defined, but easy trail. This we did, and stayed there at the Hotel Beaumont, one of the first in the world, if not the first, to be electric lighted, having been wired personally by Edison.

By this time it was apparent that, especially in the high altitudes of Colorado, I was not as good a hiker as Dirac. However, there proved to be an excellent solution as to how we could get over a rather high pass to Telluride. A horse had been trained to return home if one placed the reins over its head. So I was able to ride to the top of the pass, while Dirac walked and so I was fresh to walk with him down into Telluride. Although Dirac was the superior walker, I was the more proficient reader of a topographical map because I had taken a number of courses in geology. At the top of the pass, after looking at the map, we differed as to which gulch to go down. Fortunately, Dirac had a compass along, and it ruled in my favor.

Returning from Telluride, we left there on a train, if it can be so called, which was the other extreme from the California Limited. It was a contraption colloquially known as the 'Galloping Goose' – an old Pierce Arrow automobile mounted on flanged wheels that rode the rails, with the rear a van for mail and express, but with a few seats up front. This time we were more fortunate, and had a seat by the driver. It started to rain, and he was worried that the track might suddenly be blocked by a slide. (A local ditty was 'Little drops of water, little grains of sand make an awful lot of trouble for the Denver Rio Grande.')

Back in Ouray, we swam in its nice pool heated by natural warm springs, and the next day took a bus over the spectacular highway to Silverton, where I left Dirac as I had to return home. He tells me of an amusing episode that happened to him in Silverton after I left. To a good westerner it was inconceivable that anyone would travel in any way except in his private automobile or by horse. So one well-intentioned native seeing him walking, asked him if he had anything to eat or a place to sleep!

From Silverton to Durango there was a difficulty with Dirac's schedule. The 'mixed' (i.e. passenger and freight) train ran only tri-weekly, not on a day when he wanted to travel. However, it developed that on that day there was an extra 'sheep train' to pick up all the sheep from the mountain ranges and take them to other quarters. Dirac was allowed to ride in the caboose, but there were so many sheep loaded at different points en route that the train was late and he missed the bus from Durango to Mesa Verde. With characteristic modesty he did not try to get the conductor to send a wire ahead – I'm sure if there were one saying 'hold bus for Englishman wishing to go to Mesa Verde', it would have brought results. Instead Dirac was obliged to spend the night and take the bus the next day. At least he was able to find a hotel room in Durango, and fared better than an English woman post-doctoral fellow of mine who the year before was finally housed in a funeral home there because all the hotels were full.

EPILOGUE – 1972

A great deal has happened in the more than a third of a century which has elapsed since the period covered by this narrative. No longer are there the matrix and wave-mechanical camps in quantum mechanics; physicists instead debonairely refer to the 'Heisenberg system of representation'. The fields overlooking Lake Mendota where Dirac and I used to walk are now housing projects for married students or faculty at Wisconsin. The warbles of children have replaced those of birds. There are no longer any accommodations for making the North Circle trip in Glacier National Park, as the permanent camps were discontinued in World War II and have not been replaced. Furthermore our National Park Service seems to be less benevolent to mankind than to grizzly bears, which have become an increasing menace over the years. Hikers in the more remote areas are now advised to wear bells (like a cat!), and it would be unwise to venture off the trail as Dirac and I sometimes did to climb to an outlook point.

The United States no longer has any all-Pullman trains and I know of no train anywhere in the world today that would refuse 500 mile passengers. Travelling in Colorado lacks the local color it had in 1934. The canyon of the Gunnison where we walked is now being flooded for a power project. I visited Ouray in 1968 and learned that there are no more horses trained to return home, but their role is replaced by jeeps that can negotiate the trails. The Beaumont Hotel was boarded up, but there are many new motels in the whole San Juan area. Practically all the tracks of the narrow gauge lines of the Denver Rio Grande have been scrapped. An exception is the Silverton–Durango sector. The sheep are doubtless now handled by truck, but this line is now a historic landmark and tourist attraction with one or two ten-car trains each day during the summer. A 'Galloping Goose' is preserved in the railway museum at Golden, Colorado. As there were originally three or four of these strange animals, there is no way of knowing whether it is the one on which Dirac rode. Otherwise it should, of course, carry a plaque commemorating doubtless its most distinguished passenger.

My final paragraph might well be labelled 'Dirac and Heisenberg around the world in 1972'. I do not mean to imply that they are again travelling together, but rather that their names are universally familiar to the faculty and students in physical science everywhere. In 1929 the light of Dirac's discoveries had reached only the avant-garde, so to speak, of theoretical physicists who were cognizant of the quantum-mechanical revolution. Today he is a well-known permanent star in the history of physics.

NOTES

1. M. Born, W. Heisenberg and P. Jordan, *Z. Physik* **35**, 567 (1926).
2. P. A. M. Dirac, *Proc. Roy. Soc.* (*London*) **109**, 642 (1925); **110**, 561 (1926); **111A**, 281 (1926).
3. J. H. Van Vleck, *Helv. Phys. Acta* **41**, 1234 (1968).
4. E. L. Hill and J. H. Van Vleck, *Phys. Rev.* **32**, 250 (1928).
5. P. A. M. Dirac, *Proc. Roy. Soc.* (*London*) **A117**, 610 (1928).
6. J. H. Van Vleck, *Phys. Rev.* **31**, 587 (1928).

7. C. G. Darwin, *Proc. Roy. Soc. (London)* A115, 1; A116, 227 (1927).
8. A. Sommerfeld, p. 24 of the Report of the Sixth (1930) Solvay Conference 'Le Magnetisme'.
9. C. G. Darwin, *Proc. Roy. Soc. (London)* A118, 654 (1928).
10. P. A. M. Dirac, *Proc. Roy. Soc. (London)* A123, 714 (1929).
11. J. H. Van Vleck, 'The theory of electric and magnetic susceptibilities', pp. 322ff.
12. J. H. Van Vleck, *Phys. Rev.* 45, 405 (1934).
13. J. H. Van Vleck and A. Sherman, *Rev. Mod. Phys.* 7, 167 (1935).
14. Note added in proof: Professor Albert A. Bartlett has undertaken this project, but has been thwarted by a gap of several years in the archival files of the *Lake City Silver World*. He does find that the Sept. 8, 1934 issue of the *Silverton Standard and Miner* reports an 'extra sheep train', to which we allude on p. 14, but carries no mention of its extra rider from England or his visit to Silverton.

3

'The Golden Age of Theoretical Physics': P. A. M. Dirac's Scientific Work from 1924 to 1933†

Jagdish Mehra

In June 1968, at the Trieste Symposium on Contemporary Physics, Paul Dirac introduced Werner Heisenberg when he gave one of the evening lectures in the series 'From a life of physics'. Dirac said:

I have the best of reasons for admiring Heisenberg. He and I were young research students at the same time, about the same age, working on the same problem. Heisenberg succeeded where I failed. There was a large mass of spectroscopic data accumulated at that time and Heisenberg found the proper way of handling it. In doing so he started the golden age of theoretical physics, and for a few years after that it was easy for any second rate student to do first rate work.[1]

In this article I shall discuss Dirac's intellectual background and his contributions to theoretical physics during the years 1924–33. Dirac's scientific work in this remarkable period, in spite of his down-to-earth modesty, shows that he was himself one of the principal architects of the golden age of theoretical physics.

1. GROWING UP AND EDUCATION IN BRISTOL

Paul Adrien Maurice Dirac was born on 8 August 1902 in Bristol, England, the son of Charles Adrien Ladislas Dirac and his wife Florence Hannah Holten. He was the second of three children, having an older brother and a younger sister. His father, who was Swiss by birth, had left home for England as a young man and married there. He taught French in the Merchant Venturer's Secondary School in Bristol, the school in which Dirac also received his early education. At home, young Dirac followed the rule which his father had set, of talking to him in French in order to learn the language. Whenever Paul found that he could not express himself well in French, he would stay silent. This arrangement rather early led to a habit of reticence. The need for social

† The references in brackets in this article are to items 1 to 38 of Dirac's bibliography published in this volume.

contacts was not much emphasized in his home. Paul was an introvert and, being often silent and alone, he devoted himself to the quiet contemplation of nature.

Dirac's father appreciated the importance of a good education and encouraged him to study mathematics. Merchant Venturer's School in Bristol was a very good school, concentrating on science (mathematics, physics, and chemistry, but no biology), modern languages, and a little history and geography. In contrast to most secondary schools, Latin and the classics were not included in the curriculum; students intending to go to Oxford or Cambridge, where Latin was required for admission, would learn it as a separate subject. Most of the students going on to college from Dirac's school would go to Bristol University and pursue studies in science or engineering. Dirac himself did not like the arts side of the curriculum and considered himself lucky to have gone to this school.[2] The school shared its buildings with the engineering college of Bristol University, so that it was a secondary school in the daytime and a technical college in the evening. As a result the laboratory facilities of the college were available to the school, and Dirac even obtained some practical experience of metal work. In physics there were three hours a week of lectures with one afternoon of practical work, and the course covered heat, light and sound. Dirac's chemistry teacher believed in teaching chemistry in the modern way and introduced atoms and chemical equations very early in the course, using atomic weights rather than equivalent weights right away. In most of the subjects, which also included courses in English, French and German, Paul did not go much beyond the class, with the exception of mathematics. Mathematics, which interested Dirac most, was divided into algebra, geometry, and trigonometry. He improved and extended his knowledge by going throughly through the few books that were available to him. He did a lot of mathematical reading on his own, at a more advanced level than the rest of the class. He worked rather early through books on calculus (Edwards) and geometry (Hall and Knight). Even modern methods of non-Euclidean geometry were pursued at the school. The older students were sent off to do war work (World War I), leaving empty the higher classes, and Dirac had the advantage of being pushed into a class higher than would correspond to his age throughout his secondary school, thereby obtaining the opportunity of learning things, especially the sciences, quite early.

His elder brother had studied engineering, and Paul tended to follow in his footsteps. On completing school at the age of sixteen, Dirac entered the University of Bristol as a student of engineering. His studies were carried on in the same building as before, and even some of the teachers were the same. One of them, David Robertson, who had taught him physics at school was now his professor of electrical engineering. Robertson taught both theoretical and practical subjects, and under his influence Dirac chose to specialize in electrical engineering. Robertson was paralyzed from polio and had to get around in a wheelchair; he had organized his life very methodically and he impressed upon his students the same need for organization. Dirac learned all about electrical circuits, and Robertson performed the calculations in such a way that mathematical beauty and elegance always showed up. Interestingly enough, he did not teach electromagnetic theory or electromagnetic waves, but mainly the topics of power engineering. Dirac did learn from him some Heaviside calculus while working with linear differential equations. The emphasis was not so much on strict proofs, as on mathematical rules by which one could get the right results in some magical way.

Dirac also learned some general engineering such as the testing of materials, calculating stresses in structures, etc. It was probably this training that first gave him the idea of a delta function. 'Because when you think of loads of engineering structures, sometimes you have a distributed load and sometimes you have a concentrated load. In the two cases you have somewhat different equations. Essentially, the attempt to unify these two things leads to the delta function.'[3] At that time Dirac did not unify the treatment but he subconsciously felt the necessity of doing so. An important thing which he learned from his engineering courses was that approximations, by which one had to describe the actual situation, could also yield beautiful and satisfying mathematics. 'Previously to that I thought any kind of an approximation was really intolerable and one should just concentrate on exact equations all the time. Then I got the idea that in the actual world, all of our equations are only approximate. We must just *tend* to greater and greater accuracy. In spite of the equations being approximate, they can be beautiful.'[3]

The mathematicians at Bristol had learned about Dirac's mathematical ability from the record of his examinations at school. They hoped that he would specialize in mathematics, and they were quite disappointed when he decide to go into engineering.[4] Although he liked mathematics very much and would have liked to pursue it, he did not know that one could earn a living from pure science. The thought of pursuing an

education in mathematics and becoming a school teacher did not
appeal to him. He decided that engineering would lead to a satisfactory
career, but he found he was also not particularly good at practical work
and did very little experimental work of any kind during his studies.[5]

Dirac graduated from Bristol University in 1921 with a degree in
electrical engineering at the age of nineteen. His father sent him to
Cambridge to try for a scholarship. It was too late for an open scholarship, but on the basis of an examination he was offered a scholarship
of £70 per annum at St John's College. This sum was not enough to
support him in Cambridge, and Dirac stayed on in Bristol with his
parents. He looked around for a job in engineering, but there was a
depression on at that time and he could not find one. David Robertson
suggested to him that instead of waiting around he should do some
research work in engineering and gave him a problem on stroboscopes.
A few weeks later, however, the mathematics department at Bristol
offered Dirac free tuition to study mathematics, which he did for the
following two years. The course in mathematics at Bristol University
normally lasted three years, but with his previous training they let him
off one year. During the first year Dirac studied both pure and applied
mathematics, and in the second year he had to specialize in one of the
two. Dirac and a Miss Dent were the only two students in the honours
course in mathematics. Miss Dent was quite determined to pursue
applied mathematics, and in order that the mathematical faculty should
not have to give two sets of courses, Dirac also decided on applied
mathematics. This was his way back to science, and he never looked for
a job in engineering again. The choice had been made for him.

In mathematics at Bristol Dirac came under the influence of his
teachers Peter Fraser and H. R. Hassé.[6] Hassé taught applied mathematics, and Fraser lectured on pure mathematics. Fraser introduced him
to ideas of mathematical rigour and made the subject really attractive.
Besides teaching projective geometry, which Dirac found very interesting, Fraser gave rigorous proofs in differential and integral calculus.
This was a new experience for Dirac, because he felt, and still does
today, that when one is confident that a certain method gives the right
answer, one does not have to bother with mathematical rigour. Fraser
emphasized the geometrical approach to mathematical thinking. Dirac
learned much algebra on his own, even reading about quaternions.
Most people would accept that Dirac has shown great ability as an
algebraist in his work, but Dirac has always maintained that his own
thinking about physical problems is largely 'geometrical'.

Besides mathematics, Dirac had been interested in the theory of relativity since his days as a student of engineering. At the end of World War I in 1918, the theory of relativity produced great general excitement, and newspapers and magazines were full of articles on it. In those days Dirac himself thought about space and time and how they might be connected. He thought that one might have to rotate space and time axes together, but since at that time he knew only Euclidean space, he had to give up the problem. While still in school in Bristol Dirac attended the lectures on relativity of the philosopher C. D. Broad. Broad's lectures were more philosophical than mathematical; Dirac obtained the physical point of view from A. S. Eddington's *Space, Time and Gravitation.*[7] Dirac learned special and general relativity simultaneously on his own, and his geometrical thinking helped him in appreciating the ideas.

As a student, Dirac also took some interest in philosophy and logic. He obtained a copy of John Stuart Mill's *Logic* from the library and read it through. At that time he thought that philosophy was perhaps important, but later on he decided that it 'will never lead to important discoveries. It is just a way of thinking about discoveries which have already been made.'[8]

During all his growing up and education in Bristol, Dirac lived at home with his parents and, but for the two years when he studied in the mathematics department, he attended school and college in the same building. David Robertson and Peter Fraser, who appreciated his gifts, invited him to their homes now and then. During his engineering course in 1920, Dirac worked for about two months in the long vacation at the British Thompson–Houston Works in Rugby. He obtained some practical experience, but it was not very satisfactory. This was his first stay away from home. Thereafter he went on to Cambridge and became, within a few years, one of the world's greatest physicists.

2. STUDENT IN CAMBRIDGE

Dirac arrived in Cambridge as a research student in the autumn of 1923. A grant from the Department of Scientific and Industrial Research for work in higher mathematics, together with the 1851 Exhibition Studentship which he had won two years earlier, made it possible for Dirac to support himself in Cambridge. He had done well in his mathematical studies in Bristol, but he did not know the standards of Cambridge; and he was not yet sure whether he would be good enough for teaching in a university.

Although Dirac had been admitted to St John's College, he did not obtain immediate residence there as there were not enough rooms in the college. His private lodgings were often cold, and he used to work in the libraries. Several libraries were available in Cambridge for his use: the library of the Cambridge Philosophical Society, the University Library, the library of St John's College, and the small library at the Cavendish Laboratory. In these libraries, especially at the Cavendish, it was possible to work undisturbed at a table for a whole morning. Dirac also found the libraries useful because he always relied more on his own reading for precise and detailed information than on lectures where he just got general ideas.[9]

A subject which he studied intensively in the early days at Cambridge was Hamiltonian methods. He used Whittaker's *Analytical Dynamics*.[10] The action-angle variables were considered very important in the early 1920s in the problem of the description of non-periodic motions, as great progress had been obtained with their use earlier in the Bohr–Sommerfeld quantization of multiply periodic systems. Dirac had not heard of Bohr's theory of the atom at all in Bristol, as he was there in mathematics and did not have any contact with the physicists; also, applied mathematics, which he studied, did not go much beyond potential theory and did not include the Bohr theory of the atom. In fact, Dirac got the impression that the mathematics courses in Cambridge were considerably more advanced than the ones he had had in Bristol; not only was the level higher, but new subjects like thermodynamics and the statistical mechanics of Gibbs were also treated.

Dirac had hoped that E. Cunningham, who worked in electromagnetic theory, would become his research supervisor. He had known him since his examination for the Exhibition in 1921, and Dirac thought that electromagnetic theory was the subject he might work on. For some reason Cunningham could not accept him, and he was assigned to R. H. Fowler. Dirac would see Fowler about once a week and discuss scientific problems with him. Fowler was interested in atomic physics and statistical mechanics at that time, and he immediately put Dirac to work on the problem of dissociation under a temperature gradient, giving him the necessary literature to read and suggesting what lectures he might attend.

It was in Fowler's lectures on quantum theory that Dirac first learned about the Bohr atom; it was also from Fowler that he learned statistical mechanics, including the Boltzmann equation. He did not much appreciate the latter because the most important concept in it, the collision term, was 'not explained very well'. Dirac preferred the approach of Gibbs. In any case, Dirac published a paper on statistical mechanics in the *Proceedings of the Royal Society*, dealing with the conditions of statistical equilibrium between atoms, electrons and radiation [4]. It grew out of the problems which Dirac found suggestive in the lectures of R. H. Fowler and J. E. Jones (who later changed his name to Lennard-Jones, on marrying a Miss Lennard).

Like all theoretical physicists in Cambridge, such as Cunningham and Lennard-Jones, Fowler belonged to the mathematics department, whereas experimental physics belonged to physics or experimental philosophy. Administratively there was a close connection between pure and applied mathematics, and theoretical physics came under the latter. The students in mathematics or theoretical physics remained together for a long time, specializing only very late in the curriculum. Theory and experiment in physics were rather separate fields in Cambridge. However, Fowler's closeness to Rutherford affected their students, and Dirac, for instance, would go to the experimental colloquia at the Cavendish. There even existed a club, the '$\nabla^2 V$ Club', which consisted jointly of experimental and theoretical people. At the Cavendish theoreticians and experimentalists met frequently to listen to distinguished speakers such as Niels Bohr. The separation of the specialities applied essentially to the undergraduates, and it was Fowler who bridged the gap.

Fowler's influence was very stimulating, and he was really the centre of quantum theory in Cambridge. He was very excited with it, and his excitement was infectious. He often visited Niels Bohr's institute in Copenhagen and brought back news from there. On the experimental side, Rutherford dominated the Cavendish, where he had become Sir J. J. Thomson's successor in 1919. With brilliant students and collaborators such as J. Chadwick, J. Cockcroft, M. Oliphant, Rutherford had given the Cavendish a new dimension and created a school of experimental physics, which for a long time, was far ahead of theory. It is remarkable that Paul Dirac, who knew Rutherford so well, never did work in the theory of nuclear phenomena.

When Dirac first arrived in Cambridge, Sir J. J. Thomson was still around, and he saw him occasionally at the Cavendish. More often did he see Sir Joseph Larmor, Lucasian Professor of Mathematics in Cambridge. This chair was first occupied by Newton, and Dirac would inherit it from Larmor nine years after his arrival in Cambridge. E. Cunningham introduced Dirac to classical electromagnetic theory, and

J. E. Lennard-Jones acquainted him with statistical mechanics in greater detail. E. A. Milne, who worked on relativity and astrophysics, became Dirac's supervisor during a term when Fowler was away in Copenhagen. With his new acquaintances and experiences, Dirac's scientific horizon was already greatly enlarged in his first year at Cambridge.

Among the great men around in Cambridge in those days was Arthur Eddington who had introduced Einstein's theory of relativity to England. Eddington had succeeded Sir George Darwin as Plumian Professor of Astronomy and Experimental Philosophy in 1913. In 1918 he had prepared for the Physical Society a *Report on the Relativity Theory of Gravitation*, and in order to test the predictions of Einstein's general relativity he had led, in 1919, the successful and famous Solar Eclipse Expedition which had created a tremendous effect on the public. Eddington was famous for his investigations of the internal structure, motion, and evolution of stars. Dirac had read his book *Space, Time and Gravitation* in Bristol, and his new book *The Mathematical Theory of Relativity* made its appearance when Dirac came to Cambridge.[11] Eddington influenced Dirac, and was later himself influenced by Dirac when the wave equation of the electron showed that relativity could be included in the new quantum mechanics. Like Einstein, Eddington searched for a unified theory which would explain the fine structure constant and the proton–electron mass ratio. He sought to connect cosmological quantities such as the radius of the universe with quantum concepts.

Dirac kept up his interest in geometry at Cambridge, and in the early days there he took Max Newman's course on the geometrical aspects of general relativity. He would often attend the Saturday tea parties given by Henry Baker, Lowndean Professor of Astronomy and Geometry in Cambridge, for people who were keen on geometry and Dirac would contribute to the discussions.[12] It had become very popular at that time to work with four dimensions and projective geometry was preferred to metrical geometry. Dirac kept up with mathematics only to the extent that he needed it for his own research.

On the whole, Dirac's nature and habits did not change very much in Cambridge and his schedule remained simple. During the week he would attend four or five lectures.[13] His scientific problems, at least in the early period, he discussed only with his supervisor. Only rarely would he see other research students, except at dinner in the evening. Morning and evening were devoted to studying, with short walks in the

afternoon. Occasionally he was invited to tea. He read very little literature and never went to the theatre.[12] He preferred solitude for work and
contemplation, often going for long walks. Every Sunday he would go
for a whole day's walk, taking his lunch with him.[8] He would not
intentionally think about his work during those walks, although he
might review it now and then. Often new ideas would come to him, and
it was on one of these long walks on a Sunday that it occurred to him
that the commutators might correspond to the Poisson brackets in
classical physics.[1]

3. BRIEF APPRENTICESHIP IN RESEARCH

Dirac's first publications as a research student were in thermodynamics
and statistical mechanics, relativity and the old quantum theory. The
topics he chose reflected to some extent the influence of his supervisors,
R. H. Fowler and E. A. Milne. Fowler had become interested in
statistical mechanics and in 1922, with Charles Darwin, he had started
work on the energy partition and developed the method of steepest
descents. He applied this method to ensembles in dissociative equilibrium, and proved the validity of M. N. Saha's dissociation formula
for ionization at high temperatures.[14] Later on, he applied this theory to
stellar atmospheres.[15] Dirac had been his student for only about six
months when, on 3 March 1924, his paper on 'Dissociation under a
temperature gradient' was communicated by Fowler to the *Proceedings
of the Cambridge Philosophical Society* [1]. The problem which Dirac had
studied was the spatial change of the dissociation equilibrium. in the
presence of a temperature gradient, of a gas consisting of molecules (say
hydrogen) which can 'decay' into two similar molecules (atomic
hydrogen). By considering the equations of kinetic equilibrium, Dirac
calculated a measurable change in the concentrations when the temperature gradient is applied at the ends of a tube containing the reacting
partners. Another paper by Dirac on statistical equilibrium, the subject
of which had grown out of the lectures of Fowler and Lennard-Jones,
was communicated by Rutherford to the *Proceedings of the Royal
Society* [4]. It treated the detailed balance between atoms, electrons and
radiation, and had an astrophysical application. Using the main physical
assumption that all atomic processes are reversible, Dirac employed the
methods of statistical mechanics and relativistic kinematics to give the
general treatment of a problem which Einstein and Ehrenfest had
treated a year before.[16] He concluded that the temperature dependence
on the reaction constant follows Van't Hoff's isochore law in relativistic

systems also. Finally, using his formulae, Dirac calculated the ionization of a monatomic gas in statistical equilibrium obtaining both Saha's formula and its generalization by Fowler.

Dirac wrote two papers when he worked with Milne during Fowler's visit to Copenhagen for a term. In the first of these, which was communicated by Eddington to the *Philosophical Magazine* [2], Dirac treated a problem which had only been partially solved in Eddington's *The Mathematical Theory of Relativity*, namely the identity of 'kinematic' and 'dynamical' velocity of a single particle of arbitrary shape. If the kinematic velocity of the particle vanishes, that is, all the space coordinates are constant in time, then all components of the energy momentum tensor $T^{\mu\nu}$ are continuous, except the double time-like ones, on the surface of the tube occupied by the particle. Then, since they vanish inside the particle, the dynamical velocity is also zero.

In the second paper, which Milne communicated to the *Monthly Notices of the Royal Astronomical Society* in May 1925, Dirac worked out the Compton scattering by free electrons moving in stellar atmospheres [6]. Compton had tried to explain the observed displacement of absorption lines toward the red end of the spectrum near the limb of the sun.[17] Dirac now calculated explicitly the Doppler effect due to thermal motion of the electrons and the reduction of the average wavelength of the scattered radiation, which is produced by the fact that the incident radiation is more intense relative to electrons moving towards it. He concluded that the observations could not be explained in this way.[18]

In three papers, published in the *Proceedings of the Royal Society* and of the *Cambridge Philosophical Society*, Dirac dealt with the problems of the old quantum theory which was soon to be replaced by the new mechanics. On starting his work under Fowler, Dirac had learned about the atomic and quantum theories from A. Sommerfeld's *Atombau und Spektrallinien*,[19] and in short order he read most of the papers on quantum theory from the journals, the principal ones of which in those days were the *Proceedings of the Royal Society*, *Annalen der Physik*, and *Zeitschrift für Physik*. Quantum theory, around 1924, was in a state of flux and imminent change. One had the partially successful Bohr–Sommerfeld quantization method, which had arisen from a generalization of Bohr's original ideas and allowed one to treat systems of many degrees of freedom to a certain extent. In Sommerfeld's work the Hamilton–Jacobi theory played an important role, which Dirac studied from Whittaker's *Analytical Dynamics*. The important theoretical guiding principles of the day were Ehrenfest's adiabatic principle and Bohr's correspondence principle, which many thought would provide a clue to the new theory.[20]

In the early 1920s Louis de Broglie published his first papers in the *Comptes Rendus de l'Académie des Sciences* of Paris. In late 1923, Fowler was persuaded to communicate a report summarizing de Broglie's work to the *Philosophical Magazine*.[21] De Broglie's work dealt with Bohr orbits, statistical mechanics, and light quanta, which he treated as particles having a very small mass, and his principal result was the prediction of wave properties for all material particles. Dirac had most probably read de Broglie's papers, but he did not refer to them.[22]

In his first paper on the old quantum theory Dirac tried to relate the Doppler principle to Bohr's frequency condition [3]. Schrödinger had derived, on the basis of Einstein's light quantum hypothesis and the assumption that Bohr's frequency condition holds for all frames of reference, a 'generalized' Doppler principle, by which he could express the frequency in the new frame of reference.[23] Starting from Schrödinger's result, Dirac simplified his expression to the form of a usual Doppler term and gave a new derivation of it on the basis of a relativistic generalization of the frequency to a frequency vector, which is proportional to the derivative of a phase angle ϕ with respect to the space–time coordinate.[24] This work is typical of Dirac's early papers, and reflects the way in which he found his subjects for research. During his reading he discovered the possibility of improving certain results. He often took a widely discussed problem from the current literature, and by criticizing and extending the treatment put it on a firmer basis than before. This was the only way of approaching the problems of physics he knew at that time, and it is quite remarkable how fast his education was completed.[25] In his later publications he added the feature of giving long introductory sections in which he reported what had been done previously on a given problem either by others or himself, and he never tired of improving the logical order of presentations. This was meant as much to help the reader, especially if he did not have previous knowledge of the subject, as to satisfy Dirac's own logical and aesthetic needs.

In his note on 'The adiabatic invariance of the quantum integrals' [5], the first of two papers on the adiabatic principle, Dirac proceeded on the basis of the results obtained by J. M. Burgers, a student of Ehrenfest's, who had shown that one could deduce from classical laws the invariance of quantum integrals in adiabatic changes.[26] The only condition was that there should exist no linear relation between the frequencies of the system with integral coefficients; if not, the number of degrees of freedom of the system would be reduced. Dirac first extended Burgers' equations of adiabatic motion to general adiabatic, that is infinitely slow and regular, changes; he then derived the general condition for adiabatic invariance. Dirac obtained a very simple condition, consistent with the selection rules, under which the related quantum integrals were adiabatically invariant: the ratio of frequencies had to be different from the ratio of derivatives of the same frequencies with respect to the adiabatic parameter. His paper on adiabatic invariants showed the great progress Dirac had made in his first Cambridge year. He had learned how to handle rather complicated periodic systems with the elegant tools of the Hamilton–Jacobi theory.[27] In a second contribution he studied the effect of an adiabatically varying magnetic field on an atomic system [7]. The magnetic field exerts forces on the electron which depend on its velocity, and the change in the Hamiltonian with respect to the free Hamiltonian H is given by

$$H = H_0 + \frac{e}{2mc} |\mathbf{H}| \, p_\phi, \tag{1}$$

where p_ϕ is the component of the resultant angular momentum of the system in the direction of the field $\mathbf{H}$. The second term being a constant of motion, the quantum integrals continue to be given by the same functions of position and momentum variables; however, the momentum variable changes to

$$\mathbf{p} = m\dot{\mathbf{q}} - \frac{e}{c}\mathbf{A}, \quad [\mathbf{H} = \text{curl } \mathbf{A}]. \tag{2}$$

If the field $\mathbf{H}$ changes adiabatically, only a small term would be added to the Hamiltonian H, arising from the electric field that would be generated. However, Dirac also proved the adiabatic invariance under the action of a rapidly changing magnetic field which is symmetrical about an axis through the nucleus of an atom.

These papers of Dirac, before the advent of quantum mechanics, show his desperate attempt to obtain the new quantum theory from the adiabatic hypothesis. He also tried to introduce action-angle variables into systems which are not multiply periodic, such as the helium atom. 'It seemed to me at that time that that was the only way in which one could develop quantum theory.'[12] Even when, following the work of Heisenberg, he started contributing to the new theory, some of his old concepts had still not lost their appeal for Dirac. In no small measure on account of Fowler's influence, the problems of quantum theory had become central to Dirac's interests quite early. The questions of quantum theory were considered important by Fowler, Rutherford, and their closely knit groups of students. The quantum problem was often discussed in the mathematics societies of various Cambridge colleges, and in academic clubs such as the $\nabla^2 V$ Club and the Kapitza Club.[28] Dirac's early papers reflect his complete familiarity with the continental work on quantum theory and all that was considered important, especially in Copenhagen.

In the beginning Fowler had to exert pressure on Dirac to write up his work for publication. Words did not come easily to him, and he started making two or three successive drafts from very rough notes until he was satisfied. Strangely, it was from his early lack of desire for writing that Dirac developed his precise and concise style of communication into a model of linguistic and technical accuracy. He would always work out his ideas pretty well in his mind before writing, making only the final changes in the course of writing. He would not discuss his results or show them to anybody before they were fully fixed in his mind and written up. And then he did not like changes![29] Fowler would suggest to him where to send his papers, the more important ones going to the *Proceedings of the Royal Society*. The *Proceedings of the Cambridge Philosophical Society* had a high standing and were widely read, and Dirac published there also. Dirac also worked on numerous other problems that he did not publish.[30]

Later on when P. Kapitza and Dirac became close friends, Kapitza persuaded him to take up some experimental work. Dirac had the idea that the rotation of gases at high speeds might be used for the separation of isotopes. Using his apparatus he got 'a negligible amount of separation', but he discovered a 'thermal' effect in the process. When the gas was pumped through one pipe which branched into two others, then the gas came out of the two pipes at widely different temperatures; the cause for this effect being the viscosity of the gas, by which energy from the inner layers of the gas in the first pipe was transferred to the outer layers, and one could separate the layers of gas of different temperatures by the arrangement of the two pipes. The work on isotope separation stopped when Kapitza returned to Russia in 1934, as Dirac did not have enough enthusiasm to carry on without him. Nothing more happened to this work until World War II, when the subject was taken up again with some modified apparatus at Oxford. Even then it was not done on a major scale, as it could not compete with the diffusion method.

A joint paper of Dirac and Kapitza [36] refers to another experiment which they attempted to perform. They discussed the theoretical possibility of obtaining electron diffraction from a grating of standing light waves, but their experiment could not be performed with continuous light sources, the intensity being too low, and they proposed using an intense mercury arc source[t]. However, Dirac never worked further on that experiment; his experiment on the separation of isotopes remained his only experimental work as well as his only work, theoretical or experimental, which had some connection with nuclear physics.

4. QUANTUM MECHANICS

On 28 July 1925 W. Heisenberg spoke on the jungle of problems of the old quantum theory in a seminar entitled 'Term zoology and Zeeman botany' in Cambridge at the ninety-fourth meeting of the Kapitza Club. In private conversations Heisenberg also explained his ideas and results on quantum mechanics, which he had obtained a few weeks before his appearance in Cambridge, which were contained in his forthcoming paper on 'Quantum-theoretical re-interpretation of kinematic and mechanical relations'.[31] Fowler had most probably been present at Heisenberg's seminar and asked him to send him the proof sheets of his paper. In this paper, which laid the foundation of quantum mechanics, Heisenberg had derived a new rule for quantization by considering only those quantities which could be 'controlled in principle', that is, were measurable; as such he considered the Fourier coefficients describing the position and the corresponding momentum of a periodic (multiply periodic) system. Arranging these coefficients in quadratic schemes and using the Thomas–Kuhn sum rule, Heisenberg had derived a formal multiplication rule. From it he calculated the energy states of the harmonic and a slightly anharmonic oscillator, and the rotator, and provided the fundamental basis of various successful rules which had been obtained earlier by guessing from the spectral data.

Dirac most probably attended Heisenberg's seminar, and he himself gave a talk in the Kapitza Club one week later. 'I did not hear anything about Heisenberg's matrices, until I got a copy of the proofs from Fowler.' Fowler had received the proof-sheets of Heisenberg's paper at the beginning of September 1925, found the thing interesting, but was a bit uncertain about it and wanted to know what Dirac's reaction

[t] Freeman Dyson has pointed out to me that the Dirac–Kapitza effect has recently been observed using a laser source of light.

would be. Says Dirac, 'I have often tried to recall my earlier reaction [to Heisenberg's paper] but I cannot remember what it was. I suppose it was just some disparity between that and the Hamiltonian formalism. I was so impressed then with the Hamiltonian formalism as the basis of atomic physics, that I thought anything not connected with it would not be much good. I thought there was not much in it [Heisenberg's paper] and I put it aside for a week or so.'[3,8] Dirac first had the impression that Heisenberg was preserving more of the old theoretical structure than he was actually doing. However, 'I went back to it [Heisenberg's paper] later, and suddenly it became clear to me that it was the real thing.'[3,8] Heisenberg's idea had provided the key to the 'whole mystery'. During the following weeks Dirac tried to connect Heisenberg's matrices to the action-angle variables of the Hamilton–Jacobi theory.

I worked on it intensively starting from September 1925. I think it was just a matter of weeks before I got this idea of the Poisson brackets. During a long walk on a Sunday it occurred to me that the commutator might be the analogue of the Poisson bracket, but I did not know very well what a Poisson bracket was then. I had just read a bit about it, and forgotten most of what I had read. I wanted to check up on this idea, but I could not do so because I did not have any book at home which gave Poisson brackets, and all the libraries were closed. So, I just had to wait impatiently until Monday morning when the libraries were open and check on what Poisson bracket really was. Then I found that they would fit, but I had one impatient night of waiting.[1]

From the very beginning Dirac's clarification of the relationship between Heisenberg's variables and the classical variables made the formulation look more classical, and at the same time it very cleanly isolated the small point at which the reformulation had to make a break with the classical theory. From the quantum conditions expressed in angular variables Dirac found the correspondence between Heisenberg's commutation brackets and the classical Poisson brackets for the variables X and Y

$$XY - YX = i\hbar \sum_r \left\{ \frac{\partial X}{\partial q_r} \frac{\partial Y}{\partial p_r} - \frac{\partial Y}{\partial q_r} \frac{\partial X}{\partial p_r} \right\}, \tag{3}$$

where q_r and p_r can be regarded as the action-angle variables ω_r and $\mathcal{J}_r$.[†] Dirac was now safely back on Hamiltonian ground, and he showed his new results to Fowler who fully appreciated their importance. Fowler knew what was going on in Copenhagen and Göttingen and realized that there would be competition from these places. He thought that the

† R. Peierls has generalized quantum mechanical Poisson brackets for use in non-Hamiltonian field theories [*Proc. Roy. Soc. (London)* A **214**, 143 (1952)].

results obtained in England in this field had to be published at once, and urged the *Proceedings of the Royal Society* to give immediate priority to the publication of Dirac's paper on 'The fundamental equations of quantum mechanics' [8].

In his fundamental paper Dirac first gave a summary of Heisenberg's ideas, simplifying the mathematics and making it more elegant, in a manner similar to what Born and Jordan did independently. He developed a quantum algebra, introducing the sum and product of two quantities, the reciprocal and the square root. Turning to 'quantum differentiation', Dirac obtained a relation between the derivative of a quantity and its commutator with another quantity. From this he found the key to the relationship of the quantum commutator with classical theory, as expressed in equation (3). He could now use all the apparatus of classical theory, especially taking it for granted that the commutators were consistent with the equations of motion. 'The correspondence between the quantum and classical theories lies not so much in the limiting agreement when $h \to 0$, as in the fact that the mathematical operations in the two theories obey in many cases the same law.'[32] He immediately derived Heisenberg's quantization rules and obtained the canonical equations of motion for quantum systems. Finally, in the same paper, Dirac introduced an early form of creation and annihilation operators, pointing out their analogues in classical theory.

Dirac quickly followed this paper by another a few weeks later [9]. In it he developed the algebraic laws governing the dynamical variables, the algebra of 'q-numbers' as he now called the dynamical variables which satisfy all rules of normal numbers except that their product is not necessarily commutative. The non-commutativity of the product leads to some difficulties, for instance in defining the derivative. In order to relate q-numbers to the results of experiments, one has to represent them by c-numbers (normal numbers). For instance, the q-number X has to be represented by the Fourier terms $x(n, m)$ exp $i\omega(nm)t$, where $x(nm)$ and $\omega(nm)$ are c-numbers.[33] Dirac drew further conclusions from the Poisson brackets for more general commutators, and defined the conditions under which a set of variables is canonical. Any set of canonical variables Q and P is related to another one by a transformation, but at that time he did not attribute any great importance to this transformation. He gave detailed theorems on the operations with q-numbers, and applied the rules he had obtained to multiply periodic systems in close analogy with the old quantum rules.

Dirac's aim was to apply his scheme to the hydrogen atom. He wrote its Hamiltonian by simply replacing position and momentum variables in the classical Hamiltonian by q-numbers, and proceeded to obtain the Balmer formula. This paper was an important step forward because it showed that one could work with the formal scheme developed by Dirac earlier, getting results which were closely related to experiments. Dirac did not go as far as Pauli in completing the calculation on the hydrogen atom, but it was also not necessary because it could be done along the lines indicated by Pauli; Dirac had seen an account of Pauli's work and referred to it in a footnote in his paper.[34] Dirac had used the example of the hydrogen atom to emphasize the relationship and the difference between classical theory and quantum theory; without this application his work appeared to be rather formal and symbolic.

In his next paper, submitted in late March 1926, Dirac pursued the question of dynamical variables in quantum theory still further [10]. This was the same question which had been asked after Bohr's work on the hydrogen atom, and only a few hints as to an answer had been suggested in the work of Sommerfeld, Einstein, Planck, Schwarzschild, Epstein and Ehrenfest in the following decade.[35] The question was: What are the independent canonical variables when one has to treat an atomic system with several electrons in a central force field? Dirac worked out this problem in close analogy with the corresponding classical one, by expressing the 'geometrical' relations satisfied by the classical variables in analytic form and then obtaining the quantum variables which satisfy the same algebraic relations, of course replacing the classical Poisson brackets by quantum brackets.[36] From his calculations the various features of the splitting and the intensities of spectral lines in a magnetic field (including anomalous Zeeman effect) could be obtained in agreement with the experiment.

In a note read to the Cambridge Philosophical Society on 26 July 1926, Dirac summarized the properties of the functions of q-numbers [13]. Non-commutative algebra was a strange idea in those days, although it should not have been so because the quaternions had existed for a long time and the matrix calculus, certainly in mathematics, was used widely. In the beginning, Dirac himself did not quite realize that his q-number algebra was exactly equivalent to matrix algebra. After all, he did not like Heisenberg's matrix algebra too much. Dirac's approach using q-numbers seemed to him to be different from the rules Heisenberg had used in his first paper, as well as from the approach of Born and Jordan employing representations by matrices, taking the matrices themselves

D

as fundamental. He realized that the most important thing was the non-commutation, which had bothered Heisenberg very much in the beginning.[37] The mathematicians, of course, knew about matrix problems as well as the generalized concept of linear operators. The latter theory had been developed mainly by D. Hilbert, but Dirac independently rediscovered those aspects of it which he needed for his work. The mathematicians sought to obtain higher standards of rigour and were very concerned with detailed theorems of convergence and existence, things which did not appeal to Dirac very much. It is remarkable that Dirac followed his own mathematical route quite independently. He had started with the Hamilton–Jacobi action-angle theory and seen that attempts based on it would not lead to a satisfactory solution to atomic problems. A new rule had to be introduced, which he found in the concept of q-numbers and their algebra. 'I could turn to algebra when I had the basic ideas given. But to get the new basic ideas I worked geometrically. Once the ideas are established, one can put them into algebraic form and one can proceed to deduce consequences.'[12] In the theory of the hydrogen spectrum, working 'geometrically' meant that, given the action scheme and the classical problem, one could turn to the quantum problem by replacing ordinary c-numbers by q-numbers.

It is readily seen from this sequence of papers that Dirac worked very hard and with great concentration. In 1926, during this period of astonishing creativity, Dirac also completed his Ph.D. at Cambridge with a thesis on the principles of quantum mechanics [12].[38] He gave lectures on the new theory, including talks to the Kapitza Club and the $\nabla^2 V$ Club to which he had been elected, finding it easier and more instructive to talk about things which he had just learned 'than after a number of years, because you still remember where the difficulties are'.[3]

5. COMPLETION OF THE SCHEME

Schrödinger's paper on 'Quantization as an eigenvalue problem', the first of a series of papers that followed in quick succession establishing the framework of wave mechanics, was received by the editor of *Annalen der Physik* on 27 January 1926.[39] Schrödinger had constructed a theory which, at first sight, seemed to be quite different from the schemes developed in Göttingen and Cambridge. Based on the ideas of de Broglie, his theory employed a wave function for which he wrote down a linear equation, imposing certain boundary conditions. Schrödinger succeeded in reproducing the calculation for the spectrum of hydrogen in about three pages. The great physicists in Berlin, Planck, Einstein, and M. v. Laue, were very happy with Schrödinger's work because in it one could use continuous functions throughout, and one did not have to rely on the 'nasty and ugly' matrix mechanics and the 'complicated' and apparently 'self-contradictory' philosophy of N. Bohr. Besides, Schrödinger's calculations provided an easy interpretation in terms of classical concepts which the Göttingen people sought to avoid altogether. Just a little later, however, Schrödinger himself gave a proof that Heisenberg's matrix equations could be replaced by his differential equations, showing the equivalence of the two schemes with respect to the results they yielded.

Since Dirac had developed 'a good scheme' of his own and was pursuing its consequences, he was 'delayed' in reading Schrödinger's first article. When he finally did study it, he was a trifle annoyed because he now had to learn about another method which obviously also worked well. In contrast to the people at Göttingen, however, whose first reaction was that Schrödinger's wave function could not have any real physical meaning, Dirac had no 'philosophical' prejudice against it.[40] Writing on 'The theory of quantum mechanics' in August 1926, Dirac referred to Schrödinger's work [14]. He first mentioned the results which he had obtained earlier in attempting to solve the many-electron problem [10]. There the difficulty had arisen in finding a suitable set of 'uniformizing' dynamical variables; it was connected with the existence of an exchange phenomenon, noted for the first time by Heisenberg, arising from the fact that electrons are not distinguishable from each other.[41]

As was customary with him, Dirac first recast Schrödinger's theory in his own formalism. He noted the fact that, just as one might consider p and q as dynamical variables, one should also consider the negative energy $-E$ and the time as variables corresponding to the differential relations

$$p_r = -i\hbar \frac{\partial}{\partial q_r} \quad \text{and} \quad -E = -i\hbar \frac{\partial}{\partial t}. \tag{4}$$

He had already introduced this step a few months earlier in a paper on 'Relativity quantum mechanics with an application to Compton scattering', where he talked about 'quantum time' with a view to introducing relativity into quantum mechanics [11].[42] From equation (4) he drew two conclusions: first, that only rational integral functions of E and p have meaning; second, that one cannot multiply, in general, an equation containing the ps and E by a factor from the right-hand side. Dirac then rewrote the Schrödinger equation in the form

$$F(q_r, p_r, t, E)\psi = [H(q_r, p_r, t) - E]\psi = 0, \tag{5}$$

remarking that Heisenberg's original quantum mechanics follows from a special choice of the eigenfunctions.

In section 4 of the same paper [14] dealing with the Schrödinger equation, Dirac proceeded to make another very important contribution by giving a general treatment of systems containing several identical particles. Dirac said that if there is a system with say two electrons, and one considers two states (mn) or more accurately $(m(1), n(2))$ and $(m(2), n(1))$, which are distinguished only by the fact that in the second state the two electrons have been interchanged, then according to his and Heisenberg's scheme, one has to count the two states as one.[43] With this counting procedure, however, one cannot easily describe functions which are anti-symmetrical in the electron coordinates. The general expression for the two particle eigenfunction is

$$\psi_{mn} = a_{mn}\psi_m(1)\psi_n(2) + b_{mn}\psi_m(2)\psi_n(1). \tag{6}$$

There exist, however, only two choices for the coefficients a and b. Either

$$a_{mn} = b_{mn} \quad \text{symmetrical case (Bose–Einstein} \tag{7}$$
$$\text{statistics),}$$

or

$$a_{mn} = -b_{mn} \quad \text{anti-symmetrical case (Fermi–} \tag{8}$$
$$\text{Dirac statistics).}$$

The latter case follows from Pauli's Exclusion Principle which holds for the electrons.[44] He then went on to consider gases of free particles in a volume V, obeying either Bose–Einstein statistics or the statistics deduced from the exclusion principle. For the number N_s of particles in the sth set (having the same energy E_s), he derived

$$N_s = \frac{A_s}{\exp\left(\dfrac{\alpha + E_s}{kT}\right) + 1},$$

where

$$A_s = 2\pi V (2m)^{3/2} E_s^{1/2} \frac{dE_s}{(2\pi\hbar)^3} \tag{9}$$

and α is related to the density.

Dirac's recognition of the new statistics was antedated by the work of E. Fermi, who had obtained the same results several months earlier.[45]

'I had read Fermi's paper about Fermi statistics and forgotten it completely. When I wrote up my work on the anti-symmetric wave functions, I just did not refer to it at all. Then Fermi wrote and told me and I remembered that I had previously read about it.'[3] At the time when Dirac read Fermi's paper, it did not strike him as being important and it completely slipped his mind. A few months later he rediscovered that result, and the new statistics has since then been called 'Fermi–Dirac statistics'. In his work, Dirac went beyond Fermi and linked the two statistics to the symmetry properties of the eigenfunctions. This was a most important point which had to do with a deeper discussion of the problem of identical particles. Dirac had not pondered about the statistics until this problem became 'pretty obvious' to him. When he saw the problem, however, he immediately found the solution. In all this, the Schrödinger function obviously helped him a lot and automatically led him to consider the symmetry properties of a function describing several identical particles. Another important factor in his new considerations, the exclusion principle, had also not concerned him before at all, but when he had to decide the question whether a wave function is symmetric or anti-symmetric in the exchange of two-electron coordinates he reminded himself of Pauli's rule.

Still continuing this marvellous paper [14], Dirac developed the time-dependent perturbation theory of wave mechanics, independently of Schrödinger.[46] He applied it to an atomic system, considering the radiation field as a perturbation, and derived Einstein's expression for stimulated emission and showed the equality of its coefficient to that of absorption. He pointed out that for calculating spontaneous emission more detailed knowledge about the structure of the emitting system was needed.

Dirac applied Schrödinger's approach in the calculation of the Compton effect and derived the intensity law, which was the main result of his previous paper [11], in a more direct way [15]. The problem being essentially relativistic, he transformed the position variables x_1, x_2, x_3 and t by a linear canonical transformation which, apart from the denominator, is identical with the Lorentz transformation. He solved the wave equation in the new coordinates by separation of the variables, showing that the frequency and the intensity of the scattered radiation is smaller than the one given by classical theory. Dirac had worked on these problems for several years. He had come to a point where his method, that of uniformizing variables which he had taken over from classical mechanics, seemed to fail. Right at that moment he came across Schrödinger's work which gave him the key to the solution. Schrödinger had also given a proof of the equivalence of his wave mechanical scheme with the quantum mechanical scheme developed in Göttingen and Cambridge. Dirac had reason and occasion for expressing his appreciation to Schrödinger several times.[47]

What was now left to do except applying the new schemes to detailed problems? Actually there still remained much to do, and Dirac did it. In his celebrated paper 'On the "Anschaulichen" content of quantum-theoretic kinematics and mechanics', in which he introduced the uncertainty relations, Heisenberg did not proceed on the basis of the Schrödinger method with wave packets, which is the most simple and direct way, but used the tranformation theory of Dirac and Jordan, indicating thereby that the transformation theory belonged to the most reliable foundations of the quantum theory.

The canonical transformations had already played an important role in the formulations of matrix- and q-number mechanics. In matrix mechanics the transformations had been introduced in the three-man paper of Born, Heisenberg and Jordan, in which they had also treated the transformation matrix of the perturbed system.[49] Jordan further developed the transformation theory of infinite matrices by proving that every canonical transformation, which leaves the commutation relations invariant, can be written as

$$P = SpS^{-1}, \quad Q = SqS^{-1}, \tag{10}$$

where p, q are the old, and P, Q, the new dynamical variables.[48]

Dirac's approach to the transformation theory, including the action-angle variables, started in late 1925. Shortly thereafter, he turned to the Schrödinger equation. 'After people had established the equivalence between the matrix and the wave theories, I just studied their work and tried to improve on it in a way that I had done several times previously. I think the transformation theory came out of that.'[3] Dirac gave an account of his 'playing with equations' in a twenty-one-page paper on 'The physical interpretation of the quantum dynamics' [16]. He explained in the introduction what he meant by physical interpretation. He referred to the questions that could be answered by the quantum mechanical schemes and the physical information one could get from them. In order to do so, he pointed out, it was necessary to generalize the theory of matrix representation, in which the rows and columns refer to any set of constants of integration that commute, including the 'continuously changing' constants, and that his considerations could be regarded as a development of the work of C. Lanczos.[49]

Dirac's principal step was the introduction of the δ-function as a mathematical tool.[50] The δ-function, which is supposed to be zero everywhere in its range of definition, except at the point $x = 0$, helped to formulate matrices with continuous indices and their transformations. The decisive formulae are,

$$\int \delta(\alpha' - \alpha''') Y(\alpha''', \alpha'') d\alpha''' = Y(\alpha', \alpha'')$$

and (11)

$$\int \delta'(\alpha' - \alpha''') Y(\alpha''', \alpha'') d\alpha''' = \frac{\partial Y(\alpha', \alpha'')}{\partial \alpha'}.$$

With the help of the δ-function, Dirac was able to write the transformation equations for continuous spectra of variables. Since it follows from the quantum theory that conjugate variables, such as the position matrix $q(q'q'')$ and the corresponding momentum matrix $p(q'q'')$, do not commute, he simply had to use

$$q(q'q'') = q' \delta(q' - q'')$$

and (12)

$$p(q'q'') = -i\hbar \delta'(q' - q'').$$

In the matrix scheme of Heisenberg, Born and Jordan the transformation had to bring the Hamiltonian on the principal axis, that is to diagonalize it, or

$$S^{-1}HS = E, \quad \text{with } E \text{ diagonal.} \quad (13)$$

In the new language this equation meant

$$H\left(q, -i\hbar\frac{\partial}{\partial q}\right) S_E(q) = E S_E(q). \quad (14)$$

Here $S_E(q)$ stands for the transformation matrix which transforms the Hamiltonian to the principal axis, and at the same time keeps the commutation relations for the canonical variables. $S_E(q)$ can be thought of as belonging to a specific energy eigenvalue E_1 and is thus identical with Schrödinger's wave function $\psi_E(q)$ 'The eigenfunctions of Schrödinger's wave equations are just the transformation functions for the elements of the transformation matrix, that enable one to transform . . . to a scheme in which the Hamiltonian is a diagonal matrix.'[51] Dirac had thus derived Schrödinger's differential equation from quantum mechanics. By doing so, he closed the physical proof of the identity of all schemes in the new quantum theory.

In this paper (section 6), Dirac also arrived at some quantitative statements which came close to the uncertainty relations. By considering a function of the variables of position and momentum, Dirac showed that this function had to be averaged over the entire momentum space if the position were to be given infinitely sharply. As he was about to complete this paper in Copenhagen, he gave a seminar on it. 'Probably I did not present it well enough for them to appreciate what I had done and they still felt that they had to work a good deal on it.'[8, 52] The impact of this paper, which Dirac submitted from Copenhagen in December 1926, was great. Besides introducing the δ-function, which posed many problems for the mathematicians, Dirac had created a powerful method similar to the canonical transformations of the old Hamiltonian theory. Dirac now felt, quite justifiably, that the new scheme could indeed replace classical dynamics.

6. VISITS TO COPENHAGEN AND GÖTTINGEN

Fowler had been quite keen that Dirac should go to Copenhagen for a year, but Dirac himself was worried about going to a country where he did not know the language. He actually preferred to visit Germany because he knew a little German. He made a compromise and decided to spend about half a year in Copenhagen, and another half in Göttingen. The financial problem was not so severe, especially since he held an 1851 Exhibition Senior Research Studentship and a grant from the Department of Scientific and Industrial Research. In the autumn of 1926, Dirac left for Copenhagen, arriving there sometime during the week of 10 September.

Life in Copenhagen, especially at Bohr's Institute, was different from what it had been in Cambridge. Dirac spent most of his time in Bohr's institute, and he met new people there quite often. Of course, his method of working did not change appreciably, in that he still worked by himself. Bohr's large personality infected the place with his great enthusiasm. It was very different from Göttingen and Cambridge, where most of the work on the new quantum theory had actually been done. 'Without Bohr, I think, there would have been nothing. I was very much impressed by hearing Bohr talk. It was just a very inspiring experience to be with Bohr. The personality of Bohr impressed me very deeply.'[8, 12] Bohr was a deep thinker, and he thought about all problems. Dirac remembers, for instance, the psychological problem which Bohr posed of the two gunmen, each drawing a pistol and pointing at each other, but neither daring to shoot. The solution which Bohr worked out was that 'If you make up your mind to shoot and then shoot, that is a slower process than if you shoot in response to some external stimulus. Bohr bought some toy pistols and tried this out with various people in the institute.'[53] Bohr had also thought about the psychological problem of the stock exchange; he believed that someone who thought too much about buying and selling stocks would do worse than another who buys and sells at random.[12, 53]

At that time, in late 1926, Bohr was no longer doing active work on specific problems of quantum theory. He discussed all problems, of course, and devoted his thinking to the physical description based on the new quantum theory. He still thought much about the correspondence principle, and the idea of complementarity which was taking shape in his mind. Although quantum mechanics, as developed by the Göttingen school, Schrödinger, and Dirac, had already replaced the correspondence principle in obtaining new results, Bohr still stuck to it. Dirac thought that 'When one gets so absorbed in one idea, one sticks to it always. Just as Einstein thought that non-Euclidean geometry would be the answer to everything.'[3, 12] Dirac himself did not regard the correspondence principle as of any great importance, nor was he much influenced by the idea of complementarity. 'It does not provide you with many equations which you did not have before and I feel that the last word has not been said yet about the relationship between waves and particles. When it has been said, people's ideas of complementarity will be different.'[12] It ought to be mentioned in this connection that Dirac also differed from most people in his opinion about the uncertainty principle. He has also expressed the feeling that Planck's constant might be a derived constant rather than a fundamental one.[8, 12]

In Copenhagen, Dirac lived in 'Pension Scheck' near the town hall. The walk to the Institute along the lake shore took him about twenty minutes. On Sundays, as was his custom, he would take a long walk or excursion out in the country with Bohr, someone else, or with a large group, but very often alone. He would get to the Institute about half past nine in the morning, returning to the pension for lunch, because he took all his meals there. After lunch he would return to the Institute. On his walks to and from the Institute he had time to think about his work, living very much on his own as before.

At the Institute in Copenhagen, Dirac had occasion to listen to many people. Heisenberg was not around very often during the period he was there, but he could talk to Klein. Sometimes Ehrenfest came from Leyden for a visit. He was a man of great scientific curiosity and critical sense, and he contributed much in colloquia by getting things cleared up that were not properly presented by the speaker. 'Ehrenfest was the most useful man that one has ever had at colloquia.'[12] The wider perspective which Dirac obtained in Copenhagen helped him, not only in finishing the work on the interpretation of quantum theory, but to get started on extending the scheme of quantum mechanics.

The paper on 'The quantum theory of the emission and absorption of radiation', communicated to the *Proceedings of the Royal Society* by Bohr, treated the problem of building a relativistic quantum theory [17]. But the difficulties involved were so great that Dirac found it worthwhile to look into an approximation which was not strictly relativistic. As the total system, he considered an atom in interaction with a radiation field. In order to have a discrete number of degrees of freedom for the latter, he enclosed the system in a finite box, and decomposed the radiation into its Fourier components. Now, expanding the wave function of the interacting system (i.e. of the radiation and the atom, the electric field of the latter being approximated by a varying dipole potential) into those of the free radiation, he chose the following dynamical variables,

$$b_r = N_r^{\frac{1}{2}} e^{-i\theta_r/\hbar} \quad \text{and} \quad b_r^\dagger = N_r^{\frac{1}{2}} e^{+i\theta_r/\hbar}. \tag{15}$$

The dagger here denotes the Hermitian conjugate, N_r is the absolute square of the Fourier coefficient a_r, and θ_r is a phase variable conjugate to N_r. For the bs he assumed the commutation relations

$$b_r b_r^\dagger - b_r^\dagger b_r = 1, \tag{16}$$

all others being zero. The N_r take only integral values, larger than or equal to zero. Dirac recognized the nature of b and $b^\dagger$ as annihilation and creation operators, showing that the interaction of the atom with radiation causes transitions of photons with energy E_r into those with energy E_s. By calculating the matrix elements for these transitions, Dirac obtained Einstein's A and B coefficients as functions of the interaction potential.

A year after this first paper on 'second quantization', Jordan and Wigner developed a similar scheme for Fermi fields.[54] One might wonder why Dirac himself did not proceed in this direction: on the one hand, he wanted to deal with the radiation problem, and had therefore to apply Bose statistics; on the other, he could not as yet deal with the electron field in the same relativistic manner as with the photon field.[55]

Soon after Christmas 1926 Dirac went from Copenhagen to Göttingen. On the way he stopped in Hamburg to attend a meeting of the German Physical Society, where he learned that the experimenters were still doing much work in spectroscopy, on multiplets and their intensities. In Hamburg he also very probably met Pauli, whom he had come to know personally from Copenhagen. They always got along pretty well, especially since Pauli was able to understand Dirac's point of view more quickly than most people, and Dirac had the impression that Pauli appreciated his scientific attitude even more than Heisenberg's. It is possible that he also met Sommerfeld in Hamburg at that time, or got to know him during the latter's visit to Göttingen sometime later.

The atmosphere in Göttingen was again different from that which he had experienced in Copenhagen. There Dirac had a close and very friendly relationship with Bohr, but in Göttingen there was nobody with whom he could develop such a contact. He saw Born and Franck often, and met the mathematicians Hilbert, Courant and probably Weyl. Altogether Göttingen was a more mathematical school than Copenhagen and the physics developed there reflected this fact. Among the young people there, Dirac was often together with Oppenheimer. They had rooms in the same house (the boarding house of Günther Cario), and they often went on long walks together. He also met the Russian physicist Tamm in Göttingen, and made an expedition to the Harz Mountains with him and a few others. Dirac could continue to indulge in his favourite forms of physical exercise, walking and swimming. Later on he also did a bit of rock climbing. He could do all of these things in company, and also think about his own ideas when engaged in them.

Dirac submitted two papers for publication from Göttingen. The first one, on 'The quantum theory of dispersion', treated a problem, which in the work of Heisenberg and Kramers, had helped to pave the way for quantum mechanics [18].[56] In matrix theory this problem had been dealt with in close analogy with the classical theory. Dirac felt that the quantum theory was now so well established that one could leave behind the bridge provided by the correspondence principle and use the theory of interaction between atoms and radiation which he had just developed [17]. The difficulty in the calculation of effects concerning absorption, emission and scattering of radiation had been the choice of the electric field for the atom. Dirac concluded that the dipole field model which he had used for the atom would give consistent

results for the dispersion and resonance radiation in the first approximation; however, if one tried to calculate the width of a spectral line on
the basis of this model, one would meet with a divergent result [18].
His second paper from Göttingen was 'On the quantum mechanics of
scattering processes' [19]. Born had discussed this subject in wave
mechanics, and it had led him to the statistical interpretation of the
wave function.[57] Dirac also used the wave equation, which he had
'honestly' derived from quantum mechanics in his previous paper [16],
but he now formulated it in momentum space. He applied his method
to the absorption of radiation by atoms, thus treating an old problem
of his own rather in the spirit of Born.

From 24 to 29 October 1927, Dirac attended the fifth Conseil Solvay
of Physics on 'Electrons and photons' in Brussels, together with the
other architects of the old and the new quantum theory, such as Planck,
Einstein, Bohr, Ehrenfest, Debye, Born, Heisenberg, Kramers, Compton, de Broglie, Pauli and Schrödinger.[58] Among the prominent ones,
those missing were Sommerfeld and Jordan. Dirac contributed to the
discussions at the Solvay Conference. Following the report of Born and
Heisenberg on 'Quantum mechanics', in which the main emphasis was
on the Göttingen point of view, he mentioned the correspondence
between classical and quantum mechanics that appears by using action-
angle variables.[59] Again, in the general discussion following Bohr's
report on 'The quantum postulate', Dirac commented at length on the
essential differences between the classical and quantum descriptions of
physical processes. Quantum theory, he said, describes a state by a
time-dependent wave function ψ, which can be expanded at a given
time t_1 in a series containing wave functions ψ_n with coefficients c_n.
The wave functions ψ_n are such that they do not interfere at an instant
$t > t_1$. Now Nature makes a choice sometime later and decides in favour
of the state ψ_n with the probability $|c_n|^2$. This choice cannot be renounced
and determines the future evolution of the state. Heisenberg opposed
this point of view by asserting that there was no sense in talking about
Nature making a choice, and that it is our observation that gives us the
reduction to the eigenfunction. What Dirac called a 'choice of Nature',
Heisenberg preferred to call 'observation', showing his predilection for
the language he and Bohr had developed together.[59]

7. RELATIVITY QUANTUM MECHANICS

I remember once when I was in Copenhagen, that Bohr asked me what I was
working on and I told him I was trying to get a satisfactory relativistic theory

of the electron. And Bohr said, 'But Klein and Gordon have already done that!' That answer first rather disturbed me; Bohr seemed quite satisfied by Klein's solution, but I was not because of the negative probabilities that it led to. I just kept on with it, worrying about getting a theory which would have only positive probabilities.[12]

In 1926 Klein had obtained a relativistic equation for a scalar field by inserting quantum operators for momentum and energy in the equation

$$E^2 = p^2c^2 + m^2c^4. \tag{17}$$

The resulting equation was also independently discovered by Gordon in Hamburg, and is now referred to as the Klein–Gordon equation.[60] The difficulties which perturbed Dirac were connected with two questions. First, if one used the Klein–Gordon equation for a single particle and interpreted the expression

$$\psi^*(x) \frac{\partial \psi(x)}{\partial t} - \psi(x) \frac{\partial \psi^*(x)}{\partial t} = \rho(x) \tag{18}$$

as the probability of finding this particle at a certain place, then one could have a negative probability.[61] Secondly, Dirac had already set up the transformation theory in its general form which was a very powerful tool, and he felt that it was not only correct, but had to be preserved and brought into harmony with relativity. For achieving the latter goal, he needed an equation linear in the time.

Dirac started 'playing with the equations rather than trying to introduce the right physical idea. A great deal of my work is just playing with the equations and seeing what they give. Second quantization came out from playing with equations . . . It is my habit that I like to play about with equations, just looking for mathematical relations which maybe do not have any physical meaning at all.'[8, 12] By 'introducing the right physical ideas' Dirac meant the idea of the spin. The spin of the electron had already been introduced by Uhlenbeck and Goudsmit in 1925, to explain the doublet structure of the single electron spectra without the 'hypothesis of non-mechanical stress'; Pauli had developed the theory of the spinning electron further and described the electron by a two component wave function, which could be used for explaining the empirical spectral data using a non-relativistic Schrödinger equation.[62]

Dirac's intention was to go beyond such an approximation. A scalar product in three dimensions could be formed from Pauli's σ-matrices and the momentum, and he wanted to extend it to four-dimensional space–time. After several weeks of concentrated effort he discovered the simple solution that he could do so by generalizing the 2×2

σ-matrices to 4×4 matrices, which he called γ-matrices. From the generalization of the σ-algebra, it naturally followed that the γs should anti-commute. In his derivation of the equation, Dirac had set things up in the absence of a field [20]. The homogeneity of space and time required that the coefficients of the momenta were independent of space and time, and he obtained,

$$\left(i \sum_{\mu=1}^{4} \gamma_\mu p_\mu + m \cdot c\right)\psi = 0, \qquad (19)$$

where

$$\gamma_\mu{}^2 = 1, \quad \gamma_\mu \gamma_\nu + \gamma_\nu \gamma_\mu = 2\delta_{\mu\nu}.$$

In the same paper, received by the editor on 2 January 1928, he then introduced an arbitrary electromagnetic field and replaced the components of the four-momentum by a relativistic extension of equation (2). Finally he used his equation to describe the motion of electrons in a centrally symmetric field, giving a treatment of the hydrogen spectrum.

In his second contribution on 'The quantum theory of the electron', submitted a month after the first one, Dirac proceeded to calculate the states of the hydrogen atom in his new theory [21]. He started this paper by giving the proof that 'the change of the probability of an electron being in a given volume during a given time is equal to the probability of its having crossed the boundary. This proof . . . is necessary before one can infer that the theory will give consistent results that are invariant under a Lorentz transformation' [21].

The results of the new theory were later summarized in a lecture which Dirac presented at the Leipziger Universitätswoche in June 1928 [22]. It followed from his theory that in the alkali-spectra the electron had to have a spin of magnitude $\frac{1}{2}\hbar$. The new classification of spectra was determined by the total angular momentum, which is the sum of the internal spin and the orbital momentum. The selection rules were not changed. The new theory also yielded Sommerfeld's fine structure formula. Dirac himself had thought that 'If I got anywhere nearly right with the approximation method, I would be very happy about that. I would have been too scared myself to consider it exactly because it might have given unfortunate results that would compel the whole theory to be abandoned.'[12] However, Gordon and Darwin solved the problem explicitly.[63] In his Leipzig lecture he also mentioned the problem which had bothered him the most. If one writes the wave equation with $-e$ instead of e (the electron charge), one would expect something completely new, and he speculated that it might refer to the proton. The equation, however, did not give it. He concluded at the time that, if there were no transitions between the $+e$ and the $-e$ solutions of the wave equation, it was not too bad. In his theory, the transition probability turned out to be finite, albeit very small, being of the fourth order in (v/c), where v is the velocity of the electron. The theory could therefore only be an approximation to Nature; one probably had to change the concepts entirely, even bringing in an asymmetry of the laws between past and future.

During the next two years Dirac did not publish anything on the relativistic equation.[64] This was not only due to the fact that he was lecturing and preparing the first edition of his book, *The Principles of Quantum Mechanics* [29], published in 1930, in which he included topics such as many electron systems and quantum statistical mechanics; he knew that his relativistic theory was still imperfect, and in a paper on 'A theory of electrons and protons', he explicitly gave the explanation [25]. The wave equation had, in addition to 'solutions for which the kinetic energy of the electron is positive, an equal number of unwarranted solutions with negative kinetic energy for the electron, which appear to have no physical meaning'.[65] By examining the wave function of a negative energy solution in an electromagnetic field, Dirac found that it behaved like a particle with positive charge. But this connection would not solve the problem if one did not also have the fact that the electrons follow the exclusion principle. He could therefore assume that 'there are so many electrons in the world that the most stable states are occupied, or more accurately that *all states of negative energy are occupied except perhaps a few of small velocity*'.[66] Dirac argued that the transition of electrons from states with positive energy to those with negative energy was highly suppressed, and only the unoccupied negative states, the 'holes', could be observed. He assumed that 'the holes in the distribution of negative energy electrons are the protons. When an electron of positive energy drops into a hole and fills it up, we have an electron and proton disappearing together with emission of radiation.'[67] In a following note in the *Proceedings of the Cambridge Philosophical Society* he calculated the annihilation rate [26], and in a paper read before the British Association at Bristol on 8 September 1930, Dirac summarized his results [28]. Matter consists, he said, of 'electrons and protons', and the existence of the protons 'follows from the relativistic wave equation'. A difficulty with this interpretation remained: in his theory Dirac could calculate the transition probability for the annihilation process only under the 'approximation' that the masses of the electron and proton were equal, and the resulting amplitude was several orders of magnitude higher than that suggested by empirical evidence on 'electron–proton annihilations'.[68] In spite of this problem Dirac had faith in the essential correctness of his interpretation of the wave equation.

After Dirac's publication of the electron wave equation in 1928, many people took up its study. Schrödinger himself gave an interpretation of the spin properties of a particle as 'Zitterbewegung'.[69] Dirac knew that he had to go still further in order to make the physical interpretation consistent. 'I felt that writing this paper on the electron was not so difficult as writing the paper on the physical interpretation.'[12] There was the problem with the negative energy states: 'It was an imperfection of the theory and I didn't see what could be done about it. It was only later that I got the idea of filling up all the negative energy states.'[12] Then there were the unequal masses of the positively and negatively charged particles which existed in Nature. 'I felt right at the start that the negative energy electrons would have the same rest mass as the ordinary electrons . . . I hoped that there was some lack of symmetry somewhere which would bring in the extra mass for the positively charged ones. I was hoping that in some way the Coulomb interaction might lead to such an extra mass, but I couldn't see how it could be brought about.'[70] After Weyl's careful investigations Dirac gave up the idea that the positively charged hole was a proton.

It thus appears that we must abandon the identification of the holes with protons and must find some other interpretation for them. A hole, if there were one [in the world], would be a new kind of particle, unknown to experimental physics, having the same mass and opposite charge to an electron. We may call such a particle an anti-electron. We should not expect to find any of them in Nature, on account of the rapid rate of recombination with electrons, but if they could be produced experimentally in high vacuum they would be quite stable and amenable to observation. An encounter between two hard γ-rays (of energy of at least half a million volts) could lead to the creation simultaneously of an electron and anti-electron. This probability [of the creation of a pair] is negligible, however, with the intensities of γ-rays at present available.[71]

Then on 2 August 1932 there came along the discovery of the positron by Anderson.[72] For Dirac it meant the satisfaction that his equation predicted the situation correctly, as he had hoped. His work had also provided the first example in the history of physics where the existence of a new particle was predicted on a purely theoretical basis. Dirac himself considered much more important the fact that in his equation the spin had been incorporated so naturally, just following from the symmetry properties exhibited by the equations. At the seventh Conseil Solvay in October 1933, Dirac summarized the 'Theory of the positron' [38]. In his Nobel lecture on 12 December 1933, Dirac had the occasion of returning to this topic and predicting the existence of 'negative' protons as well.[73]

8. WORK OF A THEORETICAL PHYSICIST

We have discussed Dirac's work during the period in which the theory of atomic phenomena was completed. In the same paper [31] in which Dirac created the new physical concept of the anti-particle, he also proposed the idea of a 'magnetic monopole' which, he showed, should follow from the existence of a smallest unit of charge. Besides proposing these spectacular concepts, Dirac turned his attention very early to a systematic study of relativistic quantum field theory. He had laid its foundations in the approach based on second quantization but he proceeded with much greater care and caution than many of his contemporaries. Although he was convinced that the final theory had to be fully relativistic, he always considered the approach based on successive approximations as the most desirable. In a series of three articles on quantum field theory in 1932 [33–35], Dirac proceeded more cautiously than Heisenberg and Pauli, who, in their fundamental paper, had considered the field as a dynamical system subject to quantization.[74] Dirac preferred to keep fields and particles separate. *'The very nature of an observation requires an interplay between the field and the particles.* We cannot therefore suppose the field to be a dynamical system on the same footing as the particles. The field should appear in the theory as something more elementary and fundamental.'[75] Rosenfeld showed, however, that the relativistic quantum mechanics Dirac had devised was equivalent to that of Heisenberg and Pauli.[76] Dirac further enriched quantum field theory by generalizing the concept of state to having a non-positive norm.

Dirac returned later on to a re-examination of the concept of the aether in connection with the role of time in the scheme of relativistic quantum mechanics. This idea was loosely linked to his cosmological work in the 1930s. The more unusual considerations in these papers had to do with a change of fundamental constants, such as the fine structure constant, in the course of the evolution. Dirac also contributed later on to studies on the quantization of the gravitational field.[77]

Dirac has always been very imaginative in presenting new and unfamiliar concepts, both in mathematical methods and physical interpretation. 'I think I would have been happy if any established idea was being knocked down.'[12] Jordan once proposed that one should go on from non-commutative multiplication to non-associative multiplication, and Dirac 'rather liked that idea, but did not see how to develop it mathematically'.[12] In 1968, in a talk on his life in physics, he summarized his stand on the new problems of physics by referring to the difficulties which arise in quantum electrodynamics: 'Nature certainly does not have its basic ideas described in such a clumsy and ugly way. There is probably some very neat solution which is still to be discovered.' More specifically, he added, 'It might very well involve bringing in some new representations of the Lorentz group; by suitably thinking of equations which bring in these representations one will be led to a new electrodynamics.'[1] He has always believed 'that one should keep a completely open mind for the future . . . One should not build up one's whole philosophy as though this present quantum mechanics were the last word. If one does that, one is on very uncertain ground and one will at some future time have to change one's standpoint entirely.'[78]

In Dirac's approach to the description of Nature, the use of 'pictorial models' has always been dominant. This method always became especially important to him whenever he tried to improve upon certain ideas. In the case of the quantized singularities in the electromagnetic field, for instance, he had talked about lines connecting magnetic poles [31]. 'I did not have the same idea then that all of the Faraday lines of force should be discrete.'[3, 8] Again, when he was developing his electron equation from Schrödinger's equation or the relativistic Klein–Gordon equation, he would think of wave functions 'forming some kind of density which one could picture spread out in space'.[8, 12] This picture helped him in modifying the equation and the mathematics, bringing in the four-component wave function and matrices; the mathematics thus improved the previous picture.[79] He has, for instance, never cared for the renormalization theory of quantum electrodynamics because he has not 'found the corresponding picture for the renormalization of charge'. By 'pictures', Dirac means an imaginative way of understanding the equations independently of the approximate methods of solving the equations. In the picture vague ideas become symbols, and symbols take a life of their own, until well-defined equations can be written down. The q-numbers, which Dirac had introduced in quantum mechanics, are a case in point. 'I suppose I pictured the q-numbers as some kind of mysterious numbers which represented physical things.

It was quite a time before I appreciated that they were merely operators in Hilbert space. When I did appreciate that, I dropped the terminology of q-numbers; q-numbers were some imperfectly understood physical variables satisfying non-commutative algebra.'[3, 12] The pictures also guided Dirac in manipulating concepts beyond their formal mathematical properties. For instance, when an infinity did turn up in the equations, he thought he could represent it by a finite quantity which would approximate it.

I might often imagine that a series is cut off in such a way that the neglected terms correspond to a situation which is physically unimportant. That is one way of setting up a picture which will get rid of the infinities. Then one can try to figure out whether such a picture is relativistic or not. One can do that much better than if one worked with the mathematical infinities.[3, 12]

Dirac differs strikingly from other physicists in his approach to physical problems. Heisenberg remembers a conversation in the early days of quantum mechanics, in which Dirac expressed the belief that one must handle one problem at a time, and Heisenberg felt that a whole group of problems had to be handled together or not at all.

A good many people have Heisenberg's view, but it is just too difficult to solve a whole lot of problems together. I rather early (already in Bristol) got the idea that everything in Nature was only approximate, and that science would develop through getting continually more and more accurate approximations but would never attain complete exactness. I got that point of view through my engineering training, which I think has influenced me very much. Our present quantum theory is probably only an approximation to the improvement of the future. I think several new ideas will be needed to get out of the difficulties in present day physics and that these ideas will come one at a time with intervals of several years between them.[12, 37]

Dirac's own contributions to theoretical and mathematical physics have been outstanding by the most exacting standards. In October 1932, he succeeded Sir Joseph Larmor as Lucasian Professor of Mathematics in Cambridge. In 1933, just ten years after he had started research in theoretical physics, Dirac was awarded, together with Schrödinger, the Nobel Prize for Physics for his 'discovery of new fertile forms of the theory of atoms and for its applications'.[80] Dirac's pioneering work in quantum theory brought him all the honours a scientist could achieve: Fellow of St John's College, 1927; Fellow of the Royal Society, 1930; Hopkin's prize of the Cambridge Philosophical Society for the period 1927–30; Nobel Prize, 1933; Royal Medal, 1939; and the Copley Medal of the Royal Society, 1952. In the brief portion of his scientific life discussed in this article, the golden period from

1924 to 1933, P. A. M. Dirac had himself largely created the language of theoretical physics and become the personification of that time.

ACKNOWLEDGMENTS

During the past several years I have derived enormous pleasure from conversations with Paul Dirac at various places – in Trieste and Duino, in New York, Miami and Austin – from which I have learned much about his own work and the development of quantum mechanics and recent physics. Dirac has kindly allowed me to use quotations in this article from these conversations and other interviews with him, and I wish to express my gratitude to him.

S. Chandrasekhar has told me about the years he spent in Cambridge and the personalities he knew there. We have often discussed the work of Dirac, and I am grateful to him for these conversations.

I am especially indebted to Eugene Wigner who has given me over the years every opportunity of pursuing with him scientific, historical, and philosophical questions, including the work and personality of his 'famous brother-in-law'.

I take great pleasure in thanking Helmut Rechenberg for assisting me in my research on Dirac's scientific work, and Freeman Dyson for his invaluable critique of the manuscript of this article.

Finally, I am delighted to join the other contributors to this volume in wishing Paul Dirac a very happy birthday.

NOTES

1. The evening lectures have been collected together, in a slightly condensed form, in a pamphlet entitled *From a Life of Physics*, a special supplement of the IAEA Bulletin, 1969. Some quotations in this article, however, have been taken from the complete transcripts of the lectures of Dirac and Heisenberg.

2. Dirac read few novels, mostly in connection with his study of English. He was a slow reader and once said to Oppenheimer: 'How can you do physics and poetry at the same time? The aim of science is to make difficult things understandable in a simple way; the aim of poetry is to state simple things in an incomprehensible way. The two are incompatible.' (Reported to me by Archibald McLeish in conversations, 6 April 1966.) Another story is told by George Gamow (in *Thirty Years That Shook Physics* (New York: 1966), pp. 121–2). Dirac had read through an English translation of Dostoevsky's *Crime and Punishment*, and on being asked how he liked it, he answered: 'It is nice, but in one of the chapters the author made a mistake. He describes the Sun as rising twice on the same day.' This was Dirac's only comment on Dostoevsky's novel.

3. Archive for the History of Quantum Physics (AHQP), deposited at the libraries of the *American Philosophical Society* (Philadelphia), the *University*

of California (Berkeley), and at the *Universitets Institut for Teoretisk Fysik* (Copenhagen).

4. Dirac did not have much personal contact with his teachers, and they had to learn about him from his examinations.

5. During his engineering studies, Dirac once went to a factory to get some practical training, but he failed to please his employers there. They sent an unfavourable report about his work to his professor.

    Dirac had not thought of physics as a career, perhaps because he did not come into contact with mathematicians and physicists in Bristol, as they were housed half a mile away from the engineering building.

6. Dirac believes that Peter Fraser was the best teacher he ever had. Fraser's interests covered the entire field of pure and applied mathematics, and his pupils owed much to him, for 'he sought them out at their most formative stages and talked to them, over coffee, on country walks, or on the golf course, and shared his wisdom with them'. (See the obituary notice of Fraser by W. V. D. Hodge in *J. Lond. Math. Soc.* **34**, 111–12 (1959).)

7. A. S. Eddington, *Space, Time and Gravitation* (Cambridge: 1920). Since C. D. Broad discussed the subject largely from the philosophical point of view, Dirac had to acquaint himself with tensor calculus, and he discovered that relativity was something really new about relations between space and time.[3]

8. Conversations with Dirac in Trieste, Italy, June 1968.

9. In order to understand a piece of work thoroughly, Dirac usually put it first in his own notation.

10. E. T. Whittaker, *A Treatise on the Analytical Dynamics of Particles and Rigid Bodies* (Cambridge: editions: 1904, 1917, 1927, 1937).

11. A. S. Eddington, *The Mathematical Theory of Relativity* (Cambridge: 1923).

12. Conversations with Dirac in Miami, Florida, 28 March 1969. (Dirac told me that he was fascinated with projective geometry ever since he first learned it. He had worked out a new method in projective geometry and presented it at one of these meetings, although he never published it.)

13. Dirac never followed every word that the lecturer said, but tried to obtain the major drift of ideas and filled the gaps by reading.[3, 8]

14. M. N. Saha, *Phil. Mag.* **40**, 472, 809 (1920); R. H. Fowler, *Phil. Mag.* **45**, 1 (1923).

15. R. H. Fowler (with E. A. Milne), *Mon. Not. Roy. Astr. Soc.* **83**, 403 (1923); **84**, 499 (1924).

16. Dirac referred explicitly to W. Pauli (*Z. Physik* **18**, 272 (1923)) and A. Einstein and P. Ehrenfest (*Z. Physik* **19**, 301 (1923)). Just as these authors had done, he included stimulated emission in his treatment to obtain Planck's law.

17. A. H. Compton, *Phil. Mag.* **46**, 908 (1923).

18. The problem of multiple Compton scattering in stellar atmospheres was later on treated by S. Chandrasekhar who obtained remarkable shifts. (*Proc. Roy. Soc.* (*London*) A**192**, 508 (1947/48)).

19. The first German edition of Sommerfeld's book had been published in 1919, and its English translation appeared in 1923. However, Dirac worked through the German edition with the help of a dictionary and some German which he had acquired at school.

20. Dirac thought that the correspondence principle (N. Bohr, *Kgl. Danske Videnskab. Selskab. Skr., Nat. Mat. Afd* 8, IV, 1) was not a very accurate

statement since one could not express it by a definite mathematical equation. Fowler, together with the people in Copenhagen thought, however, that this principle could act as a guiding principle for atomic theory, because it related classical and quantum theories, in the limit as Planck's constant goes to zero. Dirac himself preferred the adiabatic principle which could be formulated 'through an equation'.

21. L. de Broglie, *Phil. Mag.* **47**, 446 (1924).

22. At the end of §5 in his paper [4], p. 594, Dirac stated: 'For the discussion of equilibrium problems, quanta of radiation cannot be regarded as very small particles of matter moving with nearly the speed of light.' The idea that light quanta might have a very small mass was mentioned by Einstein and notably occurred in de Broglie's papers.

23. E. Schrödinger, *Phys. ZS.* **23**, 301 (1922).

24. This procedure shows some similarity to that of de Broglie, but the latter's paper in *Phil. Mag.*[21] appeared later.

25. Only for his first paper, the subject of which had been suggested by Fowler, had Dirac done no reading in advance. Later on, Dirac chose the subjects of his own research.

26. J. M. Burgers, *Proc. Roy. Acad. Sci. (Amsterdam)* **20**, 613 (1918).

27. Especially noteworthy was his reduction of the problem of multiple degrees of freedom via the selection rule, arousing the hope that one might be able to handle rather complicated systems with the help of the adiabatic principle.

28. The membership of the Kapitza Club was limited to the students and collaborators of P. Kapitza, and a few other people. Dirac had been elected a member shortly before Heisenberg gave his talk on 28 July 1925. He himself spoke at the following meeting of the Club (4 August 1925), and again on 15 December 1925. Although by the time of his second seminar, he had already completed his first paper on quantum mechanics, he discussed an old idea of Duane in the 'light quantum theory of diffraction'. It is remarkable that Heisenberg in his talk on 'Termzoologie and Zeeman-botanik' at the Kapitza Club had also spoken on problems of the old quantum theory, although his discovery of the quantum mechanical scheme was already several weeks old; he was apparently not certain that the solution of the main riddle was in hand.

29. N. Bohr once remarked to Heisenberg, 'Whenever Dirac sends me a manuscript, the writing is so neat and free of corrections that merely looking at it is an aesthetic pleasure. If I suggest even minor changes, Paul becomes terribly unhappy and generally changes nothing at all.' (W. Heisenberg, *Physics and Beyond* (New York: 1971), p. 87.)

   Bohr himself used to 'construct' his papers by dictating them to his scientific collaborators or a member of the family, making corrections, writing and rewriting until at last he was satisfied. Once Dirac happened to be there when Bohr was composing a paper. Bohr paced as he dictated, went back and corrected, and so it went on for a long time. Then Dirac, during an impressive pause, said: 'Professor Bohr, when I was at school, my teacher taught me not to begin a sentence until I knew how to finish it.'

30. For instance, he was quite interested in Weyl's theory of electromagnetism and gravitation (*Sitzber. Preuss. Akad. Wiss. (Berlin)* 1918, p. 465), and made a detailed study of the field around the electron, finding that the equations required a modification that would make the charge vary with time.[12]

31. Heisenberg had started this work during a stay at the rocky island of Helgoland in June 1925, and completed it in Göttingen. He sent the manuscript to Pauli for criticism on 9 July and gave the final version to Born on 11 or 12 July. (It appeared in Z. Physik 33, 879 (1925)). Born was very impressed and wrote immediately to Einstein that Heisenberg's new idea 'seems to be rather mystical but is surely correct'. On reflection, Born soon found that Heisenberg's symbolic multiplication was nothing but the matrix calculus. Born then continued the work on 'matrix mechanics' with Jordan and Heisenberg. (M. Born, and P. Jordan, Z. Physik 34, 858 (1925), and M. Born, W. Heisenberg, and P. Jordan, Z. Physik 35, 557 (1926), later cited as 'the three-man paper'.) In the latter paper, a note in proof was added, which said: 'A paper by P. A. M. Dirac (Proc. Roy. Soc. (London) A109, 642 (1925)), which has just appeared in the meantime, independently gives some of the results contained in Part I of the present paper, together with further new conclusions to be drawn from the theory.' (For details of the creation of quantum mechanics, see B. L. van der Waerden, Sources of Quantum Mechanics (Amsterdam: 1967); reprinted as Dover paperback (New York: 1968), pp. 25ff.)

32. P. A. M. Dirac, reference [8], p. 649.

33. Dirac stated that q-numbers are more general than those which can be represented by a Fourier series. Thus, in principle, non-periodic systems could be described.

34. W. Pauli, Z. Physik 36, 336 (1926), received on 17 January 1926.

35. A. Sommerfeld, Ann. Physik (Leipzig) 51, 1 (1916); M. Planck, Verh. Deutsch. Phys. Ges. 17, 407 (1915); P. Epstein, Phys. ZS 17, 148 (1916); K. Schwarzschild, Sitzber. Preuss. Akad. Wiss. (Berlin) 1916, p. 548; A. Einstein, Verh. Deutsch. Phys. Ges. 19, 82 (1917).

36. For almost two years Dirac had pondered about this question, specifically in connection with the helium spectrum. He now realized that the key to the mystery was provided by Heisenberg, who also was the first to solve the helium problem using Schrödinger's method. (Z. Phys. 39, 499 (1926).)

37. In a talk on 'Relativity and quantum mechanics' at Austin, Texas (14 April 1970) Dirac said: 'Heisenberg was led very reluctantly to noncommutative algebra, because it was so foreign to all ideas of physicists at that time, and when it first turned up he thought there must be something wrong with his theory and tried to correct it, but he was just forced to accept it.'

38. Later on Dirac lectured on the principles of quantum mechanics. His Ph.D. thesis and lectures led to his celebrated book on quantum mechanics [29].

39. E. Schrödinger, Ann. Physik (Leipzig) 79, 36, 489, 734; 80, 437; 81, 109 (1926). De Broglie had dealt with matter-wave ideas from relativistic considerations. Schrödinger also started with a relativistic equation for the motion of an electron, but he dropped this approach when he found that his attempt did not cover the empirical data on atoms. We discuss this point later on.

40. At Cambridge, C. G. Darwin was pleased with the approach based on wave mechanics, and the experimental people were, of course, happy with the new 'understandable' concept.

41. W. Heisenberg, Z. Physik 38, 411 (1926).

42. Dirac noted there that some difficulties arise between the 'quantum energy' E and the quantum mechanical Hamiltonian, because the new

variable E ought to commute with, say, the position variable, whereas the Hamiltonian does not do so.

43. This was an old question in statistical mechanics, which led in the early 1920s to a great discussion between two groups, with Einstein and Planck advocating the indistinguishability of microscopic particles (see, e.g. M. Planck, *Z. Physik* **35**, 155 (1925)) and Ehrenfest and Schrödinger raising objections (e.g. E. Schrödinger, in *Z. Physik* **25**, 41 (1924)).

44. W. Pauli, *Z. Physik* **31**, 765 (1925).

45. Fermi presented a short note 'On the quantum mechanics of a perfect monatomic gas' to the Rome Academy on 7 February 1926 (*Rend. Lincei* **3**, 145 (1926), and an extended version was received by *Zeitschrift für Physik* on 26 March 1926 (*Z. Physik* **36**, 902 (1926)). Fermi wrote to Dirac on 25 September 1926, drawing his attention to his own earlier work. (Copy of Fermi's letter in AHQP.)

46. E. Schrödinger, *Ann. Physik (Leipzig)* **81**, 109 (1926), received 21 June 1926. See also the 'three-man paper'[31].

47. See, for example, Dirac's obituary notice of Schrödinger in *Nature* **189**, 355 (1961).

48. P. Jordan, *Z. Physik* **37**, 383; **38**, 513 (1926). Jordan proceeded to develop a complete transformation theory which he submitted in December 1926 (received 18 December) to *Zeitschrift für Physik* (**40**, 809 (1927)). Equation (10) was long believed to be correct but it is not, as there are many inequivalent representations of the commutation relations.

49. C. Lanczos, *Z. Physik* **35**, 812 (1926). See also the 'three-man paper', section 3.3 on 'continuous spectra'.

50. Although the δ-function is connected with Dirac's name, it had been introduced in physics much earlier. It was indicated by G. Kirchoff (*Sitzber. Preuss. Akad. Wiss. (Berlin)* 1882, pp. 641–69, especially p. 644), and was also suggested by O. Heaviside (*Proc. Roy. Soc. (London)* A52, 504; A54, 105 (1893)); also P. Hertz used it (in 'Statistical mechanics', in Weber-Gans' *Repertorium der Physik*, vol 1/2 (Leipzig: 1916), p. 503). Dirac gave the δ-function a sharper meaning, when he remarked (in [16], p. 625): 'Strictly, of course, $\delta(x)$ is not a proper function of x, but can be regarded only as a limit of a certain sequence of functions.' The mathematical foundation of the δ-type functions was thus initiated by Dirac's intuitive use; from the work of Hadamard, Schwartz, Gel'fand and his school, a theory of generalized functions emerged which cannot be regarded as closed even today.

51. See p. 635 of [16].

52. Heisenberg and Bohr continued their intense discussions and work until February 1927, when Heisenberg arrived at detailed quantitative conclusions (formulating the Uncertainty Principle) by analysing various 'Gedanken' experiments. Dirac's transformation theory served as the basis of Heisenberg's derivation.

53. P. A. M. Dirac, 'The Versatility of Niels Bohr', in *Niels Bohr: His Life and Work* (Amsterdam: 1967), pp. 306–9.

54. P. Jordan, and E. Wigner, *Z. Physik* **47**, 631 (1928). The first attempt to quantize the free field (oscillator) had already been made in the 'three-man paper'.

55. At the end of the paper [17], Dirac thanked Bohr for conversations. Bohr had attracted his attention to the problem of field quantization, and had raised further questions to which Dirac returned only much later.

56. See the discussion in B. L. van der Waerden's introductory essay in *Sources of Quantum Mechanics.*[31]
57. M. Born, *Z. Physik* **37**, 863; **38**, 803 (1926).
58. In November 1927, on his return to Cambridge, Dirac was elected Fellow of St John's College. He travelled to the United States and Japan during 1929–30, lecturing on quantum mechanics in Michigan and Wisconsin for five months. At about the same time Heisenberg also lectured in the United States, and both took the same boat to go to Japan. They had plenty of time to reflect upon their views about Nature and discovered that they had a great deal in common; among other things both of them had a dislike of reporters. As they were approaching Yokohama, a reporter came in the pilot boat, got on board, and wanted to interview them. He soon caught Heisenberg, but Dirac managed to evade him. Dirac and Heisenberg were standing by the railing, when the reporter came up to Heisenberg and said, 'I have searched all over the ship for Dirac, but I cannot find him.' Heisenberg did not tell him that Dirac was right there standing beside him. 'He was very loyal. What he said instead was, "If you would like to ask me some questions about Dirac, I will try to answer them." So I just stood there, looking in another direction, pretending to be a stranger and listening to Heisenberg describing me to the reporter. We had a very good time travelling around Japan.' (Dirac's introduction to Heisenberg for his talk in Trieste.[1]) After a month together in Japan they separated, Heisenberg going on to India, and Dirac returning to Cambridge via Vladivostok and Moscow.

Heisenberg once told me a charming story of this ocean voyage. Dirac would often sit at a table in a corner, sipping water or a soft drink, while Heisenberg would take part in the entertainment offered by the boat, including dancing. Once, at the end of a dance, Heisenberg returned to their table, and Dirac asked him, 'Heisenberg, why do you dance?' Heisenberg said, 'Well, if a girl is nice, I like to dance with her', and he moved on to another dance. When he returned again after some time, Dirac said, 'How do you know that a girl is "nice" *before* you have danced with her?'

59. *Electrons et Photons*, Rapports et Discussions du Cinquième Conseil de Physique (Gauthier-Villars, Paris: 1928), p. 182.
60. O. Klein, *Z. Physik* **37**, 895 (1926); W. Gordon, *Z. Physik* **40**, 117 (1926); see also V. Fock, *Z. Physik* **38**, 242; **39**, 226 (1926). In his article on 'The evolution of the physicist's picture of Nature' (*Scientific American* **208**, 45–53 (May 1963)), Dirac wrote: 'I might tell you the story I heard from Schrödinger of how, when he first got the idea for this equation [the so-called Klein–Gordon equation], he immediately applied it to the behavior of the electron in the hydrogen atom, and then he got results that did not agree with experiment. The disagreement arose because at that time it was not known that the electron has a spin. That, of course, was a great disappointment to Schrödinger, and it caused him to abandon the work for some months. Then he noticed that if he applied the theory in a more approximate way, not taking into account the refinements required by relativity, to this rough approximation, his work was in agreement with observation. He published his first paper with only this rough approximation, and that way Schrödinger's wave equation was presented to the world. Afterward, of course, when people found out how to take into account correctly the spin of the electron, the discrepancy between the results of

58   J. MEHRA

applying Schrödinger's relativistic equation and the experiments was completely cleared up.'

61. Since ψ refers to a complex Klein–Gordon field, one prefers nowadays to call $\rho(x)$ the charge density rather than the particle density.
62. G. E. Uhlenbeck and S. Goudsmit, *Naturwiss.* **13**, 953 (1925); W. Pauli, *Z. Physik* **37**, 263 (1926).
63. C. G. Darwin, *Proc. Roy. Soc. (London)* A118, 654; A120, 621 (1928); W. Gordon, *Z. Physik* **48**, 11 (1928).
64. He published two papers, one on 'The basis of statistical mechanics' [23], in which he treated quantum statistical mechanics in analogy to the classical counterpart; and the other on 'Quantum mechanics of many-electron systems' [24]. In the latter, Dirac explicitly referred to the difficulties involved when higher speed particles are present; since in the problems of chemical binding these difficulties can be avoided in a good approximation, the main features of complex atomic systems follow. He returned to related problems in two later papers: a 'Note on exchange phenomena in the Thomas atom' [24] and a 'Note on the interpretation of the density matrix in the many-electron problem' [30].
65. [25], p. 360.
66. [25], p. 362.
67. [25], p. 363.
68. J. R. Oppenheimer therefore proposed filling in all the holes of negative energies (*Phys. Rev.* **35**, 562 (1930)). By doing so, first, no transitions occur, and second, electrons and protons could be regarded as independent objects.
69. E. Schrödinger, *Sitzber. Preuss. Akad. Wiss. (Berlin)*, 418 (1930).
70. Quotations are from conversations with Dirac[8, 12]. We should recall that Einstein also became interested in this problem and tried to generalize the concept of spinors to 'semi-vectors', allowing for a different mass of the 'anti-particles'. Weyl, in the second edition of *Gruppentheorie und Quantenmechanik* (Leipzig: 1931) pleaded for equal masses (p. 234).
71. See [31], pp. 61–2. Dirac remarked later on: 'I did not realize that the probability was very much greater if you just have one γ-ray hitting a nucleus.'[12]
72. C. D. Anderson did not know about Dirac's theory when he discovered the new particle in the cloud chamber (*Phys. Rev.* **43**, 492 (1933)). For details of the story of the positron, see N. R. Hanson, *The Concept of the Positron* (Cambridge: 1963), especially chapter IX.
73. Toward the close of his Nobel lecture Dirac's words were prophetic: 'Since as far as the theory is yet definite, there is a complete and perfect symmetry between positive and negative electric charge, and if this symmetry is really fundamental in nature, it must be possible to reverse the charge on any kind of particle. The negative protons would of course be much harder to produce experimentally, since a much larger energy would be required, corresponding to the larger mass.' He considered it to be an accident that the Earth prefers negative electrons and positive protons and speculated about the existence of anti-worlds. (See [38], pp. 324–5.) Indeed, Dirac liked to speculate about more perfect and 'better' theories, including those which might explain the reason for the (fine structure) constant 1/137.[8, 12]
74. W. Heisenberg and W. Pauli, *Z. Physik* **56**, 1 (1929); **59**, 168 (1929/30).
75. See [33], p. 454.

76. L. Rosenfeld, *Z. Physik* **76**, 729 (1932). Dirac, Fock and Podolsky improved the demonstration in their joint paper [34].

77. See Dirac's bibliography in this volume, items 82, 84, and 85.

78. Dirac's talk at Austin.[37] Dirac has often expressed the need for mathematical beauty in physical theory. 'I feel that a theory, if it is correct, will be a beautiful theory, because you want the principle of beauty when you are establishing fundamental laws. Since one is working from a mathematical basis, one is guided very largely by the requirement of mathematical beauty. If the equations of physics are not mathematically beautiful that denotes an imperfection, and it means that the theory is at fault and needs improvement. There are occasions when mathematical beauty should take priority over [temporary] agreement with experiment.

'Mathematical beauty appeals to one's emotions, and the need for it is accepted as an article of faith; there is no logical reason behind it. It just seems that God constructed the Universe on the basis of beautiful mathematics and we have found that the assumption that basic ideas should be expressible in terms of beautiful mathematics is a profitable assumption to make.

'In an approximate theory the mathematics may not be beautiful. Newton's theory has some beauty, but Einstein's theory of gravitation has greater beauty. Pure mathematicians, even if they are not physicists at all, and not in the least concerned with gravitation are still interested in Einstein's equations because they find the equations beautiful. The whole idea of Minkowski space and its equations is a beautiful thing because it is connected with the Lorentz group. Also non-relativistic quantum mechanics is a beautiful theory because it is complete.

'I would consider the theory of complex variables a very beautiful theory because of the great power that one has with Cauchy integrals. The same I feel with projective geometry, but not with some other branches of mathematics, such as the theory of sets and topology.

'A beautiful theory has universality and power to predict, to interpret, to set up examples and to work with them. Once you have the fundamental laws, and you want to apply them, you don't need the principle of beauty any more, because in treating practical problems one has to take into account many details and things become messy anyway.'[8, 12]

79. Dirac does not believe that he was ever very good in carrying through an algebraic calculation without picturing what the equations mean. He always preferred geometrical methods over purely algebraic ones (including topology and set theory) which are employed extensively in modern theoretical physics.

80. From the presentation address by H. Pleijel, Chairman of the Nobel Committee for Physics (*Nobel Lectures in Physics*, 1922–41 (Amsterdam: 1965), p. 289).

4

Foundation of Quantum Field Theory

Res Jost

1. INTRODUCTION

P. A. M. Dirac's scientific production in the years 1925–8, beginning with his work on 'Fundamental equations of quantum mechanics'[6] from 7 November 1925, is hardly equalled in the history of modern physics. The development of quantum mechanics initiated by Heisenberg's epoch-making paper 'Ueber quantentheoretische Umdeutung kinematischer und mechanischer Beziehungen'[5] seems to be focused in his mind. Almost all important discoveries were made or independently also made by him. There are two notable exceptions, the discovery of the uncertainty relation by Werner Heisenberg and Erwin Schrödinger's discovery of wave mechanics. The first does not leave a mark in Dirac's papers of this period, the second one is of decisive importance already in the masterpiece of 26 August 1926 'On the theory of quantum mechanics'.[8]

From an interview in June 1961 with B. L. van der Waerden ([41], p. 41) we know how this series of papers started. Dirac says:

The first I heard of Heisenberg's new ideas was in early September, when R. H. Fowler gave me the proof sheets of Heisenberg's paper.[5] At first I could not make much of it, but after about two weeks I saw that it provided the key to the problems of quantum mechanics. I proceeded to work it out myself. I had previously learnt the transformation theory of Hamiltonian Mechanics from lectures by R. H. Fowler and from Sommerfeld's book Atombau und Spektrallinien.

Sommerfeld's wonderful book was of course at that time the bible of quantum mechanics.[†]

But Sommerfeld's book does not contain the Poisson brackets, so

---

[†] It might be amusing at this point to quote from Karl T. Compton[31] on the impact of the book in America: 'I well remember when the first copy of Sommerfeld's 'Atombau und Spektrallinien' came to America in the possession of Prof. P. W. Bridgman. Until later copies arrived, he knew no peace and enjoyed no privacy, for he was besieged by friends wanting to read the book – which he would not allow to go out of his possession.'

important for Dirac's development of the fundamental equations of quantum mechanics. He must have learnt these from R. H. Fowler. I think it is not an accident that Fowler did submit most of the papers of Dirac from this period to the Royal Society.

The reason why we elaborate so much on Hamiltonian transformation theory is that Dirac's deep affinity for analytical dynamics is still noticeable in his papers on quantum electrodynamics. He made still an unrestricted use of classical canonical transformations; a use which we hardly would consider justified in our times. Unlike more puristic thinkers, Dirac lets us participate in his discoveries: old notions which in the long run might become untenable are not immediately discarded but used as stepping stones for better ones.

2. DIRAC'S PAPER ON 'THE QUANTUM THEORY OF THE EMISSION AND ABSORPTION OF RADIATION'

This paper[10] 'By P. A. M. Dirac, St John's College, Cambridge, and Institute for Theoretical Physics Copenhagen' from 2 February 1927 contains the foundation of quantum electrodynamics and the invention of the second quantization. It is the germ out of which the quantum theory of fields developed. Its impact on Dirac's contemporaries is described by G. Wentzel[39] with the following words:

Today, the novelty and boldness of Dirac's approach to the radiation problem may be hard to appreciate. During the preceding decade it had become a tradition to think of Bohr's correspondence principle as the supreme guide in such questions, and, indeed, the efforts to formulate this principle in a quantitative fashion had led to the essential ideas preparing the eventual discovery of matrix mechanics by Heisenberg. A new aspect of the problem appeared when it became possible, by quantum-mechanical perturbation theory, to treat atomic transitions induced by given external fields . . . The transition so calculated could be interpreted as being caused by absorptive processes, but the 'reaction on the field', namely the disappearance of a photon, was not described by this theory, nor was there any possibility, in this framework, of understanding the process of spontaneous emission. Here, the correspondence principle seemed indispensable, a rather foreign element (a 'magic wand' as Sommerfeld called it) in this otherwise very coherent theory. At this point, Dirac's explanation in terms of the q-matrix [i.e. the quantized vector potential] came as a revelation.

It is true that Born, Heisenberg, and Jordan in their paper[7] 'Zur Quantenmechanik II' from 15 November 1925, chapter 4, section 3, had applied their matrix mechanics to the eigenvibrations of a string mainly in order to calculate the mean square energy fluctuations. They confirmed Einstein's famous formula[1] for the mean square fluctuation for the black body radiation:

$$\epsilon^2 = h\nu E + \frac{c^3}{8\pi\nu^2 d\nu}\frac{E^2}{V}.$$

In addition Dirac himself had calculated the reaction of an atom to an external field with his time dependent perturbation theory five months earlier. He had verified Einstein's rules[2] $B_{n\to m} = B_{m\to n}$ for the coefficients of absorption and induced emission.

However, the coefficients for spontaneous emission $A_{n\to m}$ had still to be inferred from Einstein's relation

$$A_{n\to m} = \alpha\nu^3 B_{n\to m},$$

where α by Bohr's correspondence principle equals $8\pi h/c^3$. Here, as Wentzel points out, this 'Zauberstab' ([3], p. 338 and p. 707) 'by which direct use could be made of the results of the classical wave theory in quantum theory', was still necessary. Now Dirac is going to eliminate it.

The main idea appears already in '§ 1. Introduction and summary'. The attack is of unsurpassing directness. An atom interacting with the field of radiation (inside a box) is considered. If we disregard the interaction for the moment, the Hamiltonian function is of the form

$$H = \Sigma_r E_r + H_0,$$

H_0 being the Hamiltonian of the atom alone. E_r is the energy of the rth Fourier component of the radiation field. E_r can be considered in the classical theory as the canonical momentum of the rth field oscillator. Conjugate to it is the canonical coordinate θ_r, an 'angle' variable which has to be taken modulo $1/\nu_r$ where ν_r is the frequency of the rth oscillator. The vector potential is expressible by $\{E_r, \theta_r\}$. If we replace in this expression E_r and θ_r by q-numbers satisfying the canonical commutation relations, then the vector-potential itself becomes a q-number function. This quantized vector potential is now substituted into the classical interaction of the radiation field with the atom and thus, by the addition of $\Sigma_r E_r + H_0$, the quantum mechanical Hamiltonian is obtained.

Even today it seems clear that it took a genius of the unfailing formal talent of Dirac to develop the whole formalism of quantum electrodynamics from this starting point. In fact it is even more remarkable that the fundamental notion of an operator-valued field already appears in the paper 'On the theory of quantum mechanics'.[8] We read on p. 666: 'It would appear to be possible to build up an electromagnetic theory in which the potentials of the field at a specified point x_0, y_0, z_0, t_0 in space–time are represented by matrices of constant elements that are functions of x_0, y_0, z_0, t_0.' The misleading remarks on p. 677 on spontaneous emission prove that Dirac was at that time (August 1926) certainly not yet in the possession of quantum electrodynamics.

The quantization of H leads of course to photons satisfying Einstein–Bose statistics. The total Hamiltonian can equally well be interpreted as the quantum mechanical description of photons interacting with the atom. 'There is thus a complete harmony between the wave and light-quantum descriptions of the interaction' ([10], p. 245). Dirac's theory is actually built up from the light quantum point of view.

The end of the 'Introduction and summary' contains a warning, valid even today, not to confuse the complex wave function of a single photon with the necessarily real classical radiation field.

The method of passing from one system to an ensemble of systems, satisfying Einstein–Bose statistics, has the strange name: second quantization. Dirac describes it in the next two paragraphs and this description explains its name.

Quoting from his paper,[8] he begins with the Schrödinger equation for a Hamiltonian $H = H_0 + V$ in matrix notation and in the *interaction representation*. It is well known how important the interaction representation (the name is much younger than the thing itself) was going to be for the later development of field theory. It is astonishing to recognize that, as far as Dirac is concerned, this representation precedes the Schrödinger representation. For the present purpose, however, he changes to the Schrödinger representation and writes

$$i\hbar \dot{b}_r = \Sigma_s H_{rs} b_s$$

and

$$-i\hbar \dot{b}_r{}^* = \Sigma_s b_s{}^* H_{sr}.$$

This pair of equations can be interpreted as *classical* canonical equations for the *classical* Hamiltonian function

$$F = \frac{1}{i\hbar} \Sigma_{r,s} (i\hbar b_r{}^*) H_{rs} b_s,$$

where b_s is a coordinate and $(i\hbar b_s{}^*)$ the canonically conjugate momentum. These new classical equations are now subjected to a new, a second, quantization. The probability amplitudes $\{b_r\}$ become q-numbers, satisfying the commutation relations

$$b_r b_s{}^* - b_s{}^* b_r = \delta_{rs},$$

the bs commute with each other as the b^*s do.

Real (hermitean) momenta $\{N_r\}$ and conjugate coordinates $\{\theta_r\}$ are introduced by

$$b_r = e^{-i\theta_r/\hbar} N_r^{1/2} = (N_r + 1)^{1/2} e^{-i\theta_r/\hbar}$$

$$b_r^* = N_r^{1/2} e^{i\theta_r/\hbar} = e^{i\theta_r/\hbar} (N_r + 1)^{1/2}.$$

The interpretation of b_r, b_r^*, N_r as absorption, emission, and occupation number operators for an assembly of the original systems, satisfying Einstein–Bose statistics, is well known. The verification of the correctness of this interpretation is done by Dirac with utmost care and – using the benefit of Dirac's own transformation theory[9] – a representation is chosen in which all the Ns are diagonal.

With the notation

$$H_{rs} = W_r \delta_{rs} + v_{rs},$$

corresponding to the splitting of H into $H_0 + V$, the following 'second quantized' Hamiltonian is obtained:

$$F = \Sigma_r W_r N_r + \Sigma_{rs} v_{rs} N_r^{1/2} e^{i\theta_r/\hbar} (N_s + 1)^{1/2} e^{-i\theta_s/\hbar}$$

$$= \Sigma_r W_r N_r + \Sigma_{rs} v_{rs} N_r^{1/2} (N_s + 1 - \delta_{rs})^{1/2} e^{i(\theta_r - \theta_s)/\hbar}.$$

This is the most general Hamiltonian obtainable from a single particle Hamiltonian by second quantization. The Hamiltonian F commutes with the total occupation number $N = \Sigma_r N_r$ and this will lead to certain complications in the application to the radiation field coupled to an atom. Later we shall see how Dirac resolves these complications by a stroke of genius.

He turns now – and very wisely so – to the derivation of what Fermi later was to call the 'Golden Rule No. 2' ([36], p. 142). He obtains in the first order perturbation theory the expression

$$\frac{2\pi}{\hbar} |v(W^0, \gamma'; W^0, \gamma^0)|^2 \mathcal{J}(W^0, \gamma') \, d\gamma'_1 \, d\gamma'_2 \ldots d\gamma'_{n-1}$$

for the transition probability per unit time under the influence of a perturbation V with matrix elements $v(.\;;.)$. W is the 'proper energy' and $\gamma_1, \gamma_2, \ldots \gamma_{n-1}$ are additional variables which determine the state of the system. The essential new feature of this formula is the Jacobian $\mathcal{J}$, which measures the density of final states.

After this interlude the author returns to his main theme, the application of second quantization to the emission and absorption of radiation. He starts with the Hamiltonian

$$F = H_p(\mathcal{J}) + \Sigma_r W_r N_r + \Sigma_{rs} v_{rs} N_r^{1/2} (N_s + 1 - \delta_{rs})^{1/2} e^{i(\theta_r - \theta_s)/\hbar}$$

F

H_p being the Hamiltonian for the atom. $\mathcal{J}$ stands for the quantum mechanical analogue of the classical action variables and its generalization. v_{rs} are, for the moment undetermined, 'functions of the $\mathcal{J}$s and ws', where w refers somehow to the quantum mechanical analogue of the classical angle variables. The Ns are of course the number-operators of the photons.

A Hamiltonian of this form can never describe the spontaneous emission of photons, since it conserves $\Sigma_r N_r$. There are, however, unobservable, spurious, photons, i.e. the photons of frequency 0. They are supposed to correspond to $r=0$. We can imagine that in any physical state there is an infinite number of such photons. In other words we take the limit $N_0 \to \infty$ in such a way that $v_{r0}(N_0+1)^{1/2}e^{-i\theta_0/\hbar}$ $\to v_r$ and $v_{0r}N_0^{1/2}e^{i\theta_r/\hbar} \to v_r{}^*$. According to the equations of motion, θ_0 will become constant. This limit leads to the new Hamiltonian

$$F = H_p(\mathcal{J}) + \Sigma_r W_r N_r + \Sigma_{\substack{r \neq 0}} \{v_r N_r^{1/2} e^{i\theta_r/\hbar} + v_r{}^*(N_r+1)^{1/2}e^{-i\theta_r/\hbar}\}$$
$$+ \Sigma_{\substack{r \neq 0}} \Sigma_{s \neq 0} v_{rs} N_r^{1/2}(N_s+1-\delta_{rs})^{1/2}e^{i(\theta_r-\theta_s)/\hbar}.$$

This astonishing procedure of introducing spurious quanta appears here for the first time and vaguely anticipates much more radical reinterpretations of the vacuum states.

Success with this new Hamiltonian is immediate. The probability for the emission of a photon with number r is proportional to $|v_r|^2(N_r'+1)$, the probability for its absorption $|v_r|^2 N_r'$, N_r' being the number of photons before the emission. This number is related to the spectral density ρ_r by

$$\rho_r = \frac{4\pi}{c^3} \nu^2_r N'_r (2\pi\hbar\nu_r).$$

The emission is proportional to $\rho_r + \dfrac{4\pi h}{c^3} \nu^3$, the adsorption to ρ_r.

This leads, up to a factor 2, to the Einstein relation between $A_{n \to m}$ and $B_{n \to m} = B_{m \to n}$. The factor two by which the spontaneous emission is too big compared to Einstein's comes about, as Dirac points out, 'because in the present theory either component of the incident radiation can stimulate only radiation polarized in the same way, while Einstein's theory treats the two polarization components together'.

The final paragraph contains a beautifully compact and physical derivation of the expression of the vector potential in terms of the Ns and θs together with a recalculation of Einstein's Bs in dipole approximation.

Some uncertainty is expressed concerning the scattering terms involving v_{rs} ($r \neq 0$, $s \neq 0$) in the Hamiltonian. This uncertainty was to be removed in a paper published only two months later.[11] We will come back to it.

However great Dirac's success in this work, he was himself well aware of its shortcomings. The situation in which his ideas developed, the comparision between what had been achieved and what was desirable, is so marvellously described in the 'Introduction and summary' that I cannot refrain from a quotation:

The new quantum theory [Dirac writes] based on the assumption that the dynamical variables do not obey the commutative law of multiplication, has by now been developed sufficiently to form a fairly complete theory of dynamics. One can treat mathematically the problem of any dynamical system composed of a number of particles with instantaneous forces acting between them, provided it is describable by a Hamiltonian function, and one can interpret the mathematics physically by a quite definite general method. On the other hand, hardly anything has been done up to the present on quantum electrodynamics. The questions of the correct treatment of a system in which the forces are propagated with the velocity of light instead of instantaneously, of the production of an electromagnetic field by a moving electron, and of the reaction of this field on the electron have not yet been touched. In addition, there is a serious difficulty in making the theory satisfy all the requirements of the restricted principle of relativity, since a Hamiltonian function can no longer be used. This relativity question is, of course, connected with the previous ones, and it will be impossible to answer any one question completely without at the same time answering them all. However, it appears to be possible to build up a fairly satisfactory theory of the emission of radiation and of the reaction of the radiation field on the emitting system on the basis of a kinematics and dynamics which are not strictly relativistic. This is the main object of the present paper.

3. 'THE QUANTUM THEORY OF DISPERSION'

This is the title of the paper[11] 'By P. A. M. Dirac, St John's College, Cambridge; Institute for Theoretical Physics, Göttingen.' It was received 4 April 1927. It resolves the uncertainty concerning the scattering terms v_{rs} ($r \neq 0$, $s \neq 0$) in the Hamiltonian F of the previous paper. These terms do not contribute (in first approximation) to the emission and absorption of radiation. Dirac therefore had been inclined to require them to vanish. Now he will need them in order to obtain agreement with the Kramers and Heisenberg dispersion formula.[4]

At the same time he discusses already in the introduction a very mild kind of divergence difficulty, which has nothing to do with the serious troubles of higher order calculations (which Dirac does not perform), but which is connected with the derivation of 'Golden Rule No. 2'. In this derivation one has to evaluate an integral asymptotically for $t \to \infty$:

$$\int f(u) \frac{1 - \cos ut/\hbar}{u^2} \, du = \frac{t}{\hbar} \int f\left(\frac{x\hbar}{t}\right) \frac{1 - \cos x}{x} \, dx$$

$$\sim \frac{\pi t}{\hbar} f(0).$$

f is essentially the square of a matrix element. This derivation supposes at least that the original integral exists. This, however, is not the case for radiative transitions in dipole approximation. Here we see why it was so wise to first derive the 'Golden Rule No. 2' in a general setting and only afterwards to apply it to dipole transitions: the results so obtained are not only well defined but also *correct*.

How does one obtain 'the more accurate expression for the inter-action energy'? This seems simple nowadays – because we all have learnt it from Dirac. The procedure is already outlined in the 'Introduction and summary'[10] and says: quantize the vector potential (in Coulomb gauge) and substitute it into the classical interaction. Dirac had not done this in the previous paper because, in contrast to the programme of the introduction, he develops the theory by second quantization in the particle (photon) picture. Now, however, he follows exactly the procedure of the introduction. His model is an electron moving in an external electrostatic field with a potential ϕ.

If $\varkappa$ is the vector potential then the classical Hamiltonian is

$$H = c \left\{ m^2 c^2 + \left(\mathbf{p} + \frac{e}{c}\varkappa\right)^2 \right\}^{1/2} - e\phi.$$

(Is it characteristic that Dirac starts with a relativistic Hamiltonian even though he is still unable to construct a truly relativistic theory?) The non-relativistic approximation is

$$H = H_0 + \frac{e}{mc}(\mathbf{p}, \varkappa) + \frac{e^2}{2mc^2}(\varkappa, \varkappa).$$

After quantization of the radiation field the second term describes the transitions discussed in the previous paper. The third term, how-ever, apart from a divergent diagonal term which 'may be ignored' ([11], p. 717), accounts for the scattering of a photon and the double emission and absorption of photons. Dirac is very sceptical about these double emission and double absorption processes ([11], p. 718) and is inclined to consider them to be in disagreement with a correct light-quantum theory.

It is always astonishing to realize with how great a reservation Dirac and his contemporaries accepted quantum electrodynamics despite its undisputable successes. In fact there is no theory in modern physics which has been tortured quite as much and simultaneously been so good natured as to constantly yield excellent results as quantum electrodynamics. This attitude can be traced with some physicists (but not with Dirac) to the ominous role of Sommerfeld's fine structure constant $\alpha = e^2/\hbar c$ which is dimensionless, has a value pretty close[43] to $(9/8\pi^4)$ $(\pi^5/2^45\,!)^{1/4}$ and seemed so much in need for theoretical understanding, that a quantum electrodynamics, compatible with an arbitrary value of α, appeared to be doomed (W. Pauli[22], p. 272).

The new Hamiltonian in fact yields the correct Kramers–Heisenberg dispersion formula to second order in e. This formula had been the starting point in Heisenberg's first paper on matrix mechanics. Dirac's work closes the circle and non-relativistic quantum mechanics finds its final form. The riddle of the particle-wave nature of radiation, which had so strongly motivated theoretical physics since 1900, is solved. With just pride, Dirac could write in the 'Introduction and summary':

A theory of radiation has been given by the author which rests on a more definite basis. It appears that one can treat a field of radiation as a dynamical system, whose interaction with an ordinary atomic system may be described by a Hamiltonian function . . . One finds then that the Hamiltonian for the interaction of the field with an atom is of the same form as that for the interaction of an assembly of light-quanta with the atom. There is thus a complete formal reconciliation between the wave and the light-quantum points of view.

But he adds, seemingly with some regret, the ominous sentence: 'In applying the theory to the practical working out of radiation problems one must use a perturbation method, as one cannot solve the Schrödinger equation directly.'

4. TOWARDS A GENERAL THEORY OF QUANTIZED FIELDS

Wentzel in his contribution to the memorial volume to Wolfgang Pauli[39] has given an excellent account of the history of the quantum theory of fields. It would be ridiculous to try to duplicate it. No attempt will be made here to be systematic, and still less to be complete. The interested reader is referred to Wentzel.

Dirac's method of second quantization of course attracted the full attention of theoretical physicists. The question naturally arose, is such a procedure restricted to Einstein–Bose statistics or can it be modified to yield Fermi–Dirac ([8], pp. 670ff.) statistics? The answer is astonishingly simple:[15] Dirac's commutation relations have to be changed into anti-commutation relations:

$$b_r b_s + b_s b_r = 0 \quad b_r b_s{}^* + b_s{}^* b_r = \delta_{rs}.$$

It is true that this Jordan–Wigner quantization was, at the time of its invention, a very interesting but not an unavoidable formal tool ([22], p. 198). The situation, however, changed drastically after the advent of hole theory.

Dirac's papers 'The quantum theory of the electron'[12] and the 'Quantum theory of the electron II'[13] are discussed elsewhere in this volume as well as the initial difficulties connected with the physical interpretation of the negative energy states of the electron; difficulties which with the discovery of the positron in 1932 turned into an overwhelming success of the theory. Let it be sufficient to state here that the Dirac equation for the electron (and any other particle of spin $\frac{1}{2}$) has physical meaning only in conjunction with the Jordan–Wigner (second) quantization.

Long before the discovery of the positron (three years in this period of rapid development are a long time) a decisive step forward had been made by Heisenberg and Pauli in the two papers 'Zur Quantendynamik der Wellenfelder'[16] and 'Zur Quantenmechanik der Wellenfelder II'[17] from 29 March and 7 September 1929 respectively.

The starting point of Heisenberg and Pauli is an unspecified relativistically invariant classical field theory with a scalar Lagrange density. This is supposed to be a local function of the fields and their first derivatives. By a procedure well known from classical analytical dynamics, conjugate momenta to the fields are introduced. The temporal derivatives of the fields are eliminated by a Legendre transformation. The equations of motion take a canonical form. Subjecting the fields and their conjugate momenta for equal times to the Heisenberg commutation relations yields the quantum theory corresponding to the classical theory. This procedure with its asymmetrical treatment of the time is of course very far from being manifestly relativistically invariant. It amounts to a real 'tour de force' to prove that, despite its non relativistic nature, Lorentz-invariance is not destroyed by this canonical quantization. The proof given in the first paper is so complicated that the authors – following ideas by J. von Neumann – replace it by a simpler one in their second paper.

The result is then applied to quantum electrodynamics, more specifically to a system of a finite number of Dirac electrons in interaction with the electromagnetic field. This requires some modifications. Firstly, canonical quantization for the Dirac field leads to Einstein–Bose statistics. It has to be replaced by Jordan–Wigner quantization. Secondly, and this is more annoying, the electromagnetic field does not quite fit canonical quantization, since one of the canonical momenta vanishes identically. This is due to the vanishing rest mass of the photon or, equivalently, due to the presence of the electromagnetic gauge group. Therefore quantum electrodynamics is considered in these papers as the limiting case $\epsilon = 0$ of a family of theories which depend on a parameter $\epsilon > 0$ and which do not suffer from the disease of a vanishing canonical momentum.

Heisenberg and Pauli are of course more than well aware of the shortcomings of their theory: the divergence difficulties and the problem of negative energies for the electron. The importance of the paper can, however, hardly be overestimated. It opened the road to a general theory of quantized fields and thereby prepared the tools, admittedly not perfect tools, for the Pauli–Fermi theory of β-decay and the meson theories.

The first application to an 'academic field' occurs in a paper by Pauli and Weisskopf: 'Ueber die Quantisierung der skalaren relativistischen Wellengleichung'[24] where the old Klein–Gordon equation is quantized and shown to lead to results which are physically as 'reasonable' as those of Dirac's hole theory. The question why 'Nature apparently did not make use of these particles' ([24], p. 713) why, in other words, there are no stable negatively charged Bosons 'in the world as we know it', was answered much later by Dyson and Lenard.[42]

The Heisenberg–Pauli theory did not find Dirac's approval. In the paper 'Relativistic quantum mechanics'[19] we find the remarks:

It becomes necessary then to abandon the idea of a given classical field and to have instead a field which is of dynamical significance and acts in accordance with quantum laws.

An attempt at a comprehensive theory on these lines has been made by Heisenberg and Pauli. These authors regard the field itself as a dynamical system amenable to Hamiltonian treatment and its interaction with the particles as describable by an interaction energy, so that the usual methods of Hamiltonian quantum mechanics may be applied. There are serious objections to these views, apart from the purely mathematical difficulties to which they lead. If we wish to make an observation on a system of interacting particles, the only effective method of procedure is to subject it to a field of electromagnetic radiation and see how they react. Thus the rôle of the field is to provide a means of making observations. *The very nature of an observation requires an interplay between the field and the particles.* We cannot therefore suppose the field to be a dynamical system on the same footing as the particles and thus something to be observed in the same way as the particles. The field should appear in the theory as something more elementary and fundamental.

Dirac, motivated by such general arguments, which are of a certain beauty and evoke albeit vague but fascinating associations on an intimate relation between the electromagnetic field and localization, proposes the following equations for a finite number n of charged particles in interaction with the electromagnetic field[†]:

$$\Box A = 0$$

$$i \frac{\partial \Psi}{\partial t_k} = H_k(p_k, q_k, A(t_k, \mathbf{x}_k)) \Psi.$$

A is the free quantized (four) vector potential of the electromagnetic field, $H_k(p_k, q_k, A(t_k, \mathbf{x}_k))$ the Hamiltonian for the kth particle moving in the field of the vector potential A. $(t_k, \mathbf{x}_k)$ is the position of the kth particle in space-time. $\Psi(t_1, t_2, \ldots, t_n)$ is an element of the tensor product of the particle Hilbert space with the space of states of the radiation field. In addition the following Fermi-type[18] supplementary condition on Ψ is needed in order to eliminate the unphysical degrees of freedom of the vector potential.

$$A^\nu{}_{,\nu}(t, \mathbf{x}) + c \Sigma e_k D(t - t_k, \mathbf{x} - \mathbf{x}_k) \Psi = 0,$$

e_k being the charge of the kth particle and D the famous Pauli–Jordan function.[14] The equations for Ψ are of course only compatible as long as $\partial^2 \Psi / \partial t_k \partial t_l = \partial^2 \Psi / \partial t_l \partial t_k$, or as long as

$$[H_k(p_k, q_k, A(t_k, \mathbf{x}_k)), H_l(p_l, q_l, A(t_l, \mathbf{x}_l))] = 0.$$

This, however, will be the case if

$$|\mathbf{x}_k - \mathbf{x}_l| > c|t_k - t_l|$$

due to the local commutation relations for A. This is in agreement with Bloch's interpretation of the wave function $\Psi(t_1, \mathbf{x}_1, t_2, \mathbf{x}_2; \ldots; t_n, \mathbf{x}_n)$ in the configuration-space representation. It is the probability amplitude for finding particle k at the space–time position (t_k, x_n) $(k = 1, 2, 3, \ldots, n)$.[25] Observations at these positions do not interfere if all the separations between pairs of them are space-like.

The great advantage of Dirac's new theory is its manifest relativistic

[†] In [19] he illustrates his ideas with a one-dimensional model. Equations equivalent to ours appear in [20].

invariance. However, as was demonstrated most clearly by Dirac, Fock, and Podolsky[20] this new formalism is equivalent to the Heisenberg–Pauli theory. In fact the Heisenberg–Pauli Schrödinger wave-function Φ is related to Ψ by

$$\Phi(t) = e^{-itH_0/\hbar} \Psi(t, t, \ldots, t),$$

where H_0 is the Hamiltonian of the free radiation field.

Even if Dirac did not succeed in finding a new and better theory of quantum electrodynamics, his new formalism was a great advance over previous ones. It was essential for S. Tomonaga, J. Schwinger, and F. J. Dyson for the development of renormalization theory some fifteen years later. It is in fact an illustration of the usefulness of representations, which are intermediate between the Schrödinger and the Heisenberg representations. As we have seen, they occur for the first time in Dirac's paper 'On the theory of quantum-mechanics' 1926.[8]

But also R. Feynman's Space–Time approach to quantum field theory has its origin in a paper of Dirac from this period with the title: 'The Lagrangian in quantum mechanics'.[21] Since its content does not refer to quantized fields we shall not discuss it here.

5. TROUBLES

According to Pauli ('Paul Ehrenfest'[23]), Paul Ehrenfest may have been the first to suspect a fundamental difficulty in Dirac's paper on the emission and absorption of radiation. Since, in an essential way, this theory makes use of the vector potential at the position of the electron, it corresponds to the classical theory of the point electron. It has therefore, according to Ehrenfest, to lead to an infinite self energy of the electron. And Pauli comments: 'Eine Schwierigkeit, die sich beim weiteren Ausbau der Quantenelektrodynamik in der Tat als überaus peinlich und störend erweisen sollte und bis heute ungelöst ist.'

It is true that hole theory improves the situation to some extent (and actually very much, as much later developments showed). The electron self energy is 'only' logarithmically divergent;[26] instead of quadratically. But hole theory suffered from its own additional problems. If a theory contains divergences or contradictions, anything can be expected: even an infinite self-energy of the photon despite formal gauge invariance.[29] And Dirac announced at the Solvay Conference 1933[27] a logarithmically divergent charge renormalization. But he also pointed out a fascinating *finite* correction to electrodynamics, to the polarization of the vacuum. This was the beginning of a new development which was to bear its most valuable fruits fifteen years later in renormalization theory. The preparation of finite physically meaningful results from a divergent theory is always a tricky business. It was very difficult in the mid-thirties. One reason was the asymmetric treatment of the positive and negative electrons, another one the absence of a suitable adaptation of the (manifestly relativistically invariant) many time formalism of Dirac to hole theory – and the lack of a corresponding perturbation theory. Dirac and Heisenberg discovered a practicable way to deal with the problem of vacuum polarization and the non-linearities for the electro-magnetic field, which are induced by the virtual pairs in the vacuum. We will not try to describe their methods and only remark that a frontal attack on the charge (or energy-momentum) density proved to be impossible. It is well known that the charge density in Dirac's theory of the positron is a local bilinear expression in ψ_α and $\bar\psi_\beta$ and is deter-mined by the local limit of

$$\bar\psi_{\alpha'}(x')\psi_{\alpha''}(x'') - \psi_{\alpha''}(x'')\bar\psi_{\alpha'}(x').$$

This expression is singular along the light cone $(x'-x'',\ x'-x'')=0$. It has to be corrected by an additive term independent of ψ and $\bar\psi$ but depending on the external field. This term should not destroy the conservation laws and should be so chosen that for the sum the local limit exists. Heisenberg[29] was able to solve this problem and thereby to find a unique answer for the local limit.

This procedure has its origin in the paper 'Discussion of the infinite distributions of electrons in the theory of the positron' by Dirac.[28] As a starting point, Dirac discusses the vacuum expectation value of $\bar\psi_{\alpha'}(x')\psi_{\alpha''}(x'')$ for a vanishing external field and computes for this purpose the explicit expression of the invariant Δ-functions in terms of Bessel functions for the first time. The importance for modern field theory of these functions, of which the Jordan–Pauli D-function is a special (and elementary) case, is well known.

This 1934 paper is for a long time Dirac's last contribution to hole theory and 'subtraction physics' (Pauli[30]). He leaves the main stream of quantum physics. He does not show any interest in the applications of quantum field theory to mesons and nuclear forces. He goes his own way. His reasons can be guessed. The formalism had become revoltingly complex, the basic ideas obscured. Simple problems – the vacuum itself – had become unmanageable. His feelings are probably accurately expressed on the last page of the 1947 edition of his great book *The Principles of Quantum Mechanics*[35] (p. 308):

> We have here a fundamental difficulty in quantum electrodynamics, a difficulty which has not yet been solved. It may be that the wave equation (126) has solutions which are not of the form of a power series in *e*. Such solutions have not yet been found. If they exist they are presumably very complicated. Thus even if they exist the theory would not be satisfactory, as we should require of a satisfactory theory that its equations have a simple solution for any simple physical problem, and the solution of (126) for the trivial problem of the motion of a single charged particle in the absence of any incident field of radiation has not yet been found.

We know of course that the one-particle states are trivial in axiomatic field theory. But we do not know how to specify a simple (non trivial) theory in this general framework. And if we start from a Lagrangian field theory we have the greatest troubles to construct even the vacuum state. Dirac touches here on a problem, on a source of anguish, which everybody has felt at times; what are the truly elementary phenomena which are close to the foundations of a (new and better) theory and simultaneously amenable to the experiment?

Dirac goes his own way. Was it possible that the problem of infinities should first be solved in classical relativistic physics and quantization should come afterwards? He tried this path in 'La théorie de l'électron et du champs électromagnétique'[32] and 'The physical interpretation of quantum mechanics'.[33] He was prepared to sacrifice much, too much as we know today. He did give up not only Hilbert space but also the vacuum polarization and the Coulomb interaction of charged particles created in pairs (Pauli[34]).

Maybe the notion of an electron should not be part of a classical, pre-quantum, theory at all. Possibly 'the troubles of the present quantum electrodynamics should be ascribed primarily . . . not to a fault in the general principles of quantization, but to our working from a wrong classical theory'. ('A new classical theory of electrons' Dirac[37].) His new theory contains only continuous distributions of charges. The electron and the fine structure constant were expected to be a result of quantization. Dirac devoted two additional papers to this fascinating programme.[38] The theory never reached quantization.

The troubles of quantum electrodynamics are still focal points of Dirac's thinking. In recent years his critical mind turned again to renormalization theory.[40] His criticisms are not always justified. His courage and endurance to launch himself into these complicated and intricate calculations deserves, however, admiration.

6. EPILOGUE

Three physicists above all are prominent by their contributions to quantum electrodynamics in the first third of our century: Max Planck, Albert Einstein and P. A. M. Dirac. It could be a highly attractive and important contribution to the history of science to analyse and to compare the motives, the methods, and the personalities of these eminent scientists.

Max Planck: His aim was to demonstrate the irreversible nature of the black body radiation. He did not try to discover a new formula for the spectral density of the black body radiation, because he firmly believed in Wien's radiation law. But he also believed firmly in absolute irreversibility and in the deterministic interpretation of Clausius' principle for the entropy increase. He wanted to harmonize this principle with Maxwell's theory and he failed. Instead, under the pressure of experimental evidence against Wien's law, he discovered the quantum of action and his radiation formula.

Albert Einstein: He never got tired of analysing the physical meaning of Planck's formula. With his wonderful honesty and complete freedom from prejudice he penetrated deeper than anybody else into the strange rules which determine the processes of emission and absorption of electromagnetic radiation by atomic systems. He deeply distrusted the formalism and in spite of this he prepared the way for the most revolutionary formal development of modern physics, for quantum mechanics.

And P. A. M. Dirac who finally found the solution which harmonized the apparent contradictions, mastered the formalism and for this reason never permitted his formulae to drown the essential physical content.

Looking back on the history of quantum electrodynamics, we realise how different and manifold are the talents and personalities of the men who are needed to bring us closer to the solution of the great problems which Nature presents to us.

REFERENCES
1. A. Einstein, *Phys. Z.* **10**, 185 (1909).
2. A. Einstein, *Phys. Z.* **18**, 121 (1917).
3. A. Sommerfeld, *Atombau und Spektrallinien*, 3. *Auflage* (Braunschweig: 1922).
4. H. A. Kramers and W. Heisenberg, *Z. Phys.* **31**, 681 (1925).

5. W. Heisenberg, *Z. Physik* **33**, 879 (1925).
6. P. A. M. Dirac, *Proc. Roy. Soc. (London)* A109, 642–53 (1925).
7. M. Born, W. Heisenberg, and P. Jordan, *Z. Physik* **35**, 557 (1926).
8. P. A. M. Dirac, *Proc. Roy. Soc. (London)* A112, 661–77 (1926).
9. P. A. M. Dirac, *Proc. Roy. Soc. (London)* A113, 621–41 (1927).
10. P. A. M. Dirac, *Proc. Roy. Soc. (London)* A114, 243–65 (1927).
11. P. A. M. Dirac, *Proc. Roy. Soc. (London)* A114, 710–18 (1927).
12. P. A. M. Dirac, *Proc. Roy. Soc. (London)* A117, 610–24 (1928).
13. P. A. M. Dirac, *Proc. Roy. Soc. (London)* A118, 351–61 (1928).
14. P. Jordan and W. Pauli, *Z. Physik* **47**, 151 (1928).
15. P. Jordan and E. Wigner, *Z. Physik* **47**, 631 (1928).
16. W. Heisenberg and W. Pauli, *Z. Physik* **56**, 1 (1929).
17. W. Heisenberg and W. Pauli, *Z. Physik* **59**, 168 (1930).
18. E. Fermi, *R. C. Accad. Lincei* **9**, 881 (1929); **12**, 431 (1930).
19. P. A. M. Dirac, *Proc. Roy. Soc. (London)* A136, 453–64 (1932).
20. P. A. M. Dirac, V. A. Fock and Boris Podolsky, *Phys. Z. Sowjet* **2**, 468–79 (1932).
21. P. A. M. Dirac, *Phys. Z. Sowjet* **3**, 64–72 (1933).
22. W. Pauli, *Die allgemeinen Prinzipien der Wellenmechanik, Handbuch der Physik* 2nd ed. vol. 24.1 (Berlin: 1933).
23. W. Pauli, *Naturwiss.* **21**, 841 (1933).
24. W. Pauli and V. Weisskopf, *Helv. Phys. Acta* **7**, 709 (1934).
25. F. Bloch, *Phys. Z. Sowjet* **5**, 301 (1934).
26. V. Weisskopf, *Z. Physik* **89**, 27 (1934); **90**, 817 (1934).
27. P. A. M. Dirac, *Théorie du positron* in *noyaux atomiques*, 7ième conseil de physique Solvay (Paris: 1934).
28. P. A. M. Dirac, *Proc. Cambridge Phil. Soc.* **30**, 150–63 (1934).
29. W. Heisenberg, *Z. Physik* **90**, 209 (1934).
30. W. Pauli and M. E. Rose, *Phys. Rev.* **49**, 462 (1936).
31. Karl T. Compton, *Nature* **139**, 229 (1937).
32. P. A. M. Dirac, *Ann. Inst. H. Poincaré* **9** (2), 13–49 (1939).
33. P. A. M. Dirac, *Proc. Roy. Soc. (London)* A180, 1–40 (1942).
34. W. Pauli, *Rev. Mod. Phys.* **15**, 175 (1943). *Helv. Phys. Acta* **19**, 234 (1946).
35. P. A. M. Dirac, *The Principles of Quantum Mechanics*, 3rd ed. (Clarendon Press, Oxford: 1947).
36. E. Fermi, *Nuclear Physics* (University of Chicago Press: 1950).
37. P. A. M. Dirac, *Proc. Roy. Soc. (London)* A209, 291–96 (1951).
38. P. A. M. Dirac, *Proc. Roy. Soc. (London)* A212, 330–9 (1951); A223, 438–45 (1954).
39. G. Wentzel, 'Quantum theory of fields (until 1947)' in *Theoretical Physics in the Twentieth Century*, M. Fierz and V. Weisskopf eds. (Interscience Publishers, New York: 1960).
40. P. A. M. Dirac, *Lectures on Quantum Field Theory* (Yeshiva University, New York: 1966).
41. B. L. v.d. Waerden ed., *Sources of Quantum Mechanics* (North-Holland Publishing Co.: 1967).
42. F. J. Dyson and A. Lenard, *J. Math. Phys.* **7**, 423 (1967); **9**, 698 (1968).
43. A. Wyler, *C.R. Acad. Sci. Paris* **269** A, 741 (1969); *Phys. Today* **24**, 17 (1971).

5

The Early History of the Theory of the Electron: 1897–1947[†]

A. Pais

1

At the present time, the phenomena of particles and their interactions appear highly complex. We are in the midst of a struggle for simplicity. This has occurred before. In an earlier period as well, the first fifty years in the development toward a theory of the electron and its interactions, models were improvised to cope with a situation which seemed not only complicated but often, fortunately, even paradoxical. An attempt will be made here to relate some of the history of the questions raised and the models used in those days.

In the middle of this period, a momentous contribution – the discovery of the wave equation of the spinning electron – stands out as simple and direct as its author, Dirac. This essay is dedicated to him with respect and in remembrance of many discussions, walks and wood-chopping expeditions.

What were the general problems which theoreticians faced at that time? In the first quarter of this century, still the classical period, electrons were pictured as small but finite bodies, and so the question of divergences was not as yet vital. Rather, the main question became how to reconcile such a particle picture with special relativity theory.

With the advent of quantum mechanics, and especially quantum field theory, the nature of the questions changed, but it took time to realize to how great an extent. In essence, these were the problems considered central in the next twenty years.

(*a*) Since conventional quantum field theory is predicated on a point model of particles or (to put it better) on the notion of local fields and local interactions, the existence of infinite self-energies and related quantities became a burning issue. Modifications, either of physical or of formal nature, were sought in the hope to obtain a finite theory.

[†] Work supported in part by the U.S. Atomic Energy Commission under Contract Number AT(30-1)-4204.

(*b*) Should one cure the classical theory first and then quantize or quantize first, cure later? The former view was often taken on the grounds that a point model theory of the electron is already divergent on the classical level. The alternative was based on the consideration that in any event the very process of quantization changes the nature of the singularities.

(*c*) Since the days of Lorentz it was held plausible that the mass of the electron was entirely electromagnetic in origin. This was referred to as a 'unitary' theory; (the use of this terminology will hopefully not create confusion). The third main question of the period was: is the theory to be unitary or dualistic, is the electron mass purely electro-magnetic or partly mechanical? On this subject, too, Dirac was to make incisive comments.

It is well known how, later on, decisive progress was made in regard to these questions: in the quantum electrodynamics of the electron, it has proved to be extremely fruitful to bypass the questions (*a*) and (*c*) by means of the renormalization program. Question (*b*) has ceased to exist (see section 5). To the accuracy of present experiments, there is agreement with the quantum electrodynamics of the electron. However, neither these developments, since 1947, nor the new complexity, as forecast in that same year by the discovery of the μ-meson are part of the present story.

At several points below, reference will be made to models and to comparisons with experiment both of which have been known for many years now to be invalid. It is not the purpose, clearly, to draw facile analogies from this in regard to the present predicaments. It is even less the intent to make light of past efforts. On the contrary, the advantage of hindsight makes it seem even more stirring to recall, once in a while, struggles such as these. From conversations and late night reading I have tried to piece together some of these developments which mark the immediate pre-history of particle physics.

2

In 1842, Stokes treated several examples of the slow motion of a solid body through an incompressible fluid.[1] He showed that the motion of the body can be described as if it had a modified mass. No wonder then that the notion of electromagnetic mass was already introduced in 1881, by J. J. Thomson[2] – well before his experiments on the existence and properties of cathode-ray corpuscles[3] – in a theoretical study of the slow motion of a charged sphere.

The hydrodynamical analogy was also emphasized by Lorentz in his 1906 Columbia University lectures.[4] However, said Lorentz, there is an important distinction. In hydrodynamics, if we were forever unable to take a solid ball out of the fluid, we would have no means to separate its 'effective mass' into its 'true' (mechanical) and 'induced' (hydrodynamical) parts. Now (Lorentz continues) 'it is very important that, in the experimental investigation of the motion of the electron, we can go one step further. This is due to the fact that the electromagnetic mass is not a constant but increases with the velocity.'

At that point, Lorentz was referring to the calculations made by Abraham[5] for the rigid extended electron which gave the following result for the electrons' electromagnetic energy E_{elm} and momentum P_{elm}

$$E_{elm} = \frac{e^2}{2a}\left(\frac{1}{\beta} \lg \frac{1+\beta}{1-\beta} - 1\right) \simeq \frac{e^2}{2a} + \tfrac{1}{2}\mu v^2 + \ldots$$

$$P_{elm} = \frac{e^2}{2ac\beta}\left(\frac{1+\beta^2}{2\beta} \lg \frac{1+\beta}{1-\beta} - 1\right) \simeq \mu v + \ldots, \tag{1}$$

where $\beta = v/c$, $\mu = 2e^2/3ac^2$, a is the radius. Experiments by Kaufmann in 1902 had led to the conclusion 'that, within the errors, the measurements are adequately represented by the Abraham formula, so that one may consider the mass of the electron as a purely electromagnetic mass'.[6]

Such was the situation when Lorentz' 1904 paper appeared in which the electron is treated as having finite extent but is subject to Lorentz contraction.[7] This yields a velocity dependence distinct from equation (1), namely

$$E_{elm} = \frac{\mu_0 c^2}{\sqrt{(1-\beta^2)}}\left(1 + \frac{\beta^2}{3}\right) \simeq \mu_0 c^2 + \tfrac{1}{2}\mu_1 v^2 + \ldots$$

$$P_{elm} = \frac{\mu v}{\sqrt{(1-\beta^2)}} \simeq \mu v + \ldots, \tag{2}$$

where $\mu_0 = 3\mu/4$, $\mu_1 = 5\mu/4$, while μ is as in equation (1). These relations are transparent if one considers the electromagnetic energy momentum tensor density $T_{\mu\nu}$ and notes, first that ('0' refers to the rest system)

$$E_{elm} = \int T_{44} d\mathbf{x} = \frac{\int T_{44}(0)d\mathbf{x}_0 - \beta^2 \int T_{11}(0)d\mathbf{x}_0}{c^2\sqrt{(1-\beta^2)}},$$

$$P_{elm} = \frac{-i}{c} \int T_{14} d\mathbf{x} = \frac{\int T_{44}(0)d\mathbf{x}_0 - \int T_{11}(0)d\mathbf{x}_0}{c^2\sqrt{(1-\beta^2)}} v,$$

secondly, that $T_{\mu\nu}$ is traceless and that the rest-system is spatially isotropic. Since the forces on the system do not balance, $\int T_{\mu 4} d\mathbf{x}$ does not transform like a four-vector.

G

This development took place just prior to the advent of the relativity theory in 1905. After Einstein, unfettered by detailed model arguments, had given his energy-momentum relation for a free particle, it was clear that something had to be done to modify the equations (2). We shall return to this point presently, after having noted first a further confrontation with experiment.

The new theoretical developments stimulated Kaufmann to redo his experiments, with the result (1906) that 'the measurements are incompatible with the Lorentz–Einstein postulate.[6] The Abraham equation and the Bucherer equation represent the observations equally well . . .'[8]

Throughout the years it has happened remarkably often that early weak interaction experiments designed to answer a new question proved to be incorrect. The experiments under discussion (done with β-decay electrons in parallel electric and magnetic fields) were no exception. For the present account, these pioneering measurements are of particular interest because of varied impacts they had on the theoretical thinking of that time. Lorentz said of them: '. . . it will be best to admit Kaufmann's conclusion, or hypothesis if we prefer so to call it, that the negative electrons have no material mass at all. This is certainly one of the most important results of modern physics . . .'[9] (It is my impression that, even as times changed, Lorentz remained predisposed to the unitary origin of the electron mass.) Moreover it caused Lorentz to express himself with much reservation about his own relativistic considerations:[10] the experiments 'are decidedly unfavourable to the idea of a contraction, such as I attempted to work out. Yet, though it seems very likely that we shall have to relinquish it altogether, it is, I think, worthwhile looking into it somewhat more closely . . .'

Kaufmann had challenged not only Lorentz, but also Einstein. But, in 1907, Einstein took a different view of the situation:[11]

Herr Kaufmann has determined the relation between [electric and magnetic deflection] of β-rays with admirable care . . . using an independent method, Herr Planck obtained results which fully agree with [the computations of] Kaufmann . . . It is further to be noted that the theories of Abraham and Bucherer yield curves which fit the observed curve considerably better than the curve obtained from relativity theory. However, in my opinion, these theories should be ascribed a rather small probability because their basic postulates concerning the mass of the moving electron are not made plausible by theoretical systems which encompass wider complexes of phenomena.

Einstein's opinion prevailed as the experiments improved.

Let us now return to the reaction which Einstein's theory had on those concerned with its implications for models of the electron. Lorentz put it as follows:[12] 'one is naturally led to ascribe to the electron another kind of energy which is just sufficient to supplement the electromagnetic energy found by us to the amount required by the relativity theory. Poincaré is the father of these considerations on the constitution of the electron.'

Poincaré, in his independent search for what amounts to an equilibrium model of the electron, had made the following proposal.[13] (Fermi's earliest papers deal with same subject.)[13] Write the first of equations (2) as

$$E_{\text{elm}} = \frac{\mu c^2}{\sqrt{(1 - \beta^2)}} - PV,$$

$$P = 3\mu c^2 / 16\pi a^3, \quad V = (4\pi a^3/3)\sqrt{(1 - \beta^2)}.$$

V is the volume of the electron; P is a pressure. If one could eliminate the $-PV$ term, one would have the desired energy–momentum relation. Poincaré suggested to add the term $\rho P g_{\mu\nu}$ to $T_{\mu\nu}$, where $\rho = 1$ inside the electron and vanishes elsewhere. This cancels the $-PV$ term in E_{elm}, in all reference frames; and it does not contribute to P_{elm}. In any event, relativity theory eliminated the possibility to distinguish between mechanical and electromagnetic mass contributions by velocity dependences. So it is to this day.

Lorentz took due note of Poincaré's remark, but was not quite satisfied since he found that the Poincaré-equilibrium is not stable against deformations.[14] (Complications arising in the study of other than uniform motion are not considered here.) Lorentz' deep awareness of the difficulties in understanding what 'is' an electron is perhaps best conveyed in his own words:[15]

Notwithstanding all this, it would, in my opinion, be quite legitimate to maintain the hypothesis of the contracting electrons, if by its means we could really make some progress in the understanding of the phenomena. In speculating about the structure of these minute particles we must not forget that there may be many possibilities not dreamt of at present; it may well be that other internal forces serve to ensure the stability of the system, and perhaps, after all, we are wholly on the wrong track when we apply to the parts of the electron our ordinary notion of force.

This last view was also expressed by Frenkel in 1925:[16]

The inner equilibrium of an extended electron becomes . . . an insoluble puzzle from the point of view of electrodynamics. I hold this puzzle (and the questions related to it) to be a scholastic problem. It has come about by an uncritical application to the elementary parts of matter (electrons) of a principle of division, which when applied to composite systems (atoms etc.) just led to these very 'smallest' parts.

The electrons are not only indivisible physically, but also geometrically. They have no extension in space at all. Inner forces between the elements of an electron do not exist because such elements are not available. The electromagnetic interpretation of the mass is thus eliminated; along with that all those difficulties disappear which are connected with the determination of exact equations of motion of an electron on the basis of the Lorentz theory.

It is interesting to reflect that Frenkel's motivation for a reconsideration of some aspects of the theory stemmed from his observation of an apparent difficulty in connection with nuclear properties:[17] Consider the He-nucleus and make the following assumptions. (1) It is built up out of four protons and two electrons, the prevailing view at that time. (2) The respective sizes of proton and electron are e^2/Mc^2 and e^2/mc^2, corresponding to the purely electromagnetic origin of their respective masses M and m. (3) The binding energy is due to Coulomb attraction. Then the binding energy must be less than $16mc^2$ which is too small by a factor ~ 4. This is then interpreted as a fundamental difficulty for the electromagnetic theory of mass.

Such classical model considerations must at no time have caused more trouble than in the days just before the beginnings of quantum mechanics. The situation appears to have truly come to a head with the discovery of the spin of the electron.

G. E. Uhlenbeck has told me of two visits he made to Lorentz. The first one, at the instigation of Ehrenfest, was to tell Lorentz about the spin. The second one was at Lorentz' request. On that occasion Lorentz, always kind and courteous, told Uhlenbeck that he well appreciated the prospects which the spin held out for a better understanding of the spectra; but also that new problems appeared to be raised by the existence of an intrinsic angular momentum of such magnitude. (To Lorentz, classical spinning motion was no new concern.)[18] One can well imagine – such problems would appear as surface velocities of the order of $137c$.[19]

For some months after the publication by Uhlenbeck and Goudsmit, two factors of two continued to haunt the best physicists. It was at once clear that the gyromagnetic ratio g of the electron had to be equal to two in order to account quantitatively for the anomalous Zeeman effect. Such a g-value was unfamiliar. For a while it appeared intriguing that the gyromagnetic ratio of a non-relativistic spinning solid sphere is equal to two[20] – in good agreement with experiment. Simultaneously, a paradox appeared: it seemed that the fine structure splitting was off by a factor two. This was cleared up by Thomas' remark that the product of two pure Lorentz transformations is not in general a pure Lorentz transformation,[21] an observation which at that time amazed even Einstein (so Uhlenbeck tells me). Thus all was now consistent at least.

Among the several problems which the spin raised in those days, an apparent paradox noted by Rasetti and Fermi is well worth restating.[22] They reason as follows. Consider an electron with radius a and magnetic moment μ. It has a magnetic energy $\sim \mu^2/a^3$. Since a is bound to be small, equate the magnetic (rather than the electrostatic) energy to mc^2. This then gives a radius $\sim \alpha^{-2/3} \cdot e^2/mc^2 \sim 10^{-12}$ cm. Now, they continue, the radius of a good sized nucleus is known to be $\sim 10^{-12}$ cm. But such a nucleus contains a considerable number of electrons. How do these fit in?

The authors state at once that the paradox may well indicate an illegitimacy of the electron model. Understandably, their concluding remarks indicate nevertheless an acute sense of discomfort.

On his second visit, Lorentz handed Uhlenbeck a sheaf of papers with calculations dealing with the rotation of an electron which orbits around a nucleus. This work was to constitute the last communication made by the grand master of classical electron theory, to the Como Conference in September 1927.[23] In it, he stresses the great complexity of the problem and he notes that to his approximation the results are numerically different from those of Thomas. Lorentz died soon thereafter, on 4 February 1928.

Two days earlier a paper had been received by the Royal Society of London. It contained the relativistic quantum theory of the Zeeman effect. It was written by Dirac.

3

Obviously, Dirac was well aware of the need to incorporate properly the spin in the quantum mechanical description. In the first of his two papers on the Dirac equation he starts out in fact by noting that there are discrepancies which 'consist of "duplexity" phenomena, the observed number of stationary states being twice the number given by the theory'.[24] And, certainly, Pauli's treatment of the spin with the help of his σ-matrices was a source of inspiration. Yet, Dirac's foremost motivation was a different one, namely to extend the general transformation theory of quantum mechanics to the relativistic domain. It is of no concern here that subsequent insights in the theory of quantized fields were to demonstrate that the Klein–Gordon equation is less problematic

in this respect than was thought at the time. Dirac's search was for a
positive definite one particle probability density with the right covariance and conservation properties.

Spectacular though the advance was, new problems arose immediately
which were not resolved without considerable struggle. Much speculation arose at that time to the effect that the negative energy states might
be associated with the proton. For this and for another reason, a comment by Weyl is of interest:[25] 'It is plausible to anticipate that, of the
two pairs of components of the Dirac quantity, one belongs to the
electron, the other to the proton. Further, two conservation laws of
electricity will have to appear, which state (after quantization) that the
number of electrons as well as of protons remain constant. To these
conservation laws must correspond a twofold gauge invariance, involving
two arbitrary functions.' Here then is the earliest version of the conservation law of baryons, involving all baryons known in 1929.

However, said Dirac, 'one cannot simply assert that a negative
energy electron *is* a proton'.[26] He stated several paradoxes to which such
an interpretation would lead, such as transitions which violate electric
charge conservation. His next and major step forward was the hole
theory interpretation: 'Let us assume that there are so many electrons
in the world that all the most stable states are occupied, or, more
accurately, that all the states of negative energy are occupied except
perhaps a few of small velocity . . .'[26] Noting that these holes behave
like positively charged particles, 'we are therefore led to the assumption
that the holes in the distribution of negative energy electrons are the
protons'.

The mass-dissymmetry troubled him at once; for some time it was
believed that interaction effects might account for that. In a calculation
of the proton-electron annihilation he tentatively identified the m in his
equation as some mean mass.[27] So did Tamm,[28] who was quick to
notice the difficulty of an impossibly small life time of atoms, a remark
also made by Oppenheimer.[29] The latter noted further that the Thomson
scattering limit had to be the same for particle and anti-particle, in
conflict with the mass-dissymmetry.

Weyl attacked the problem with general symmetry arguments and
concluded that:

however attractive this idea [the hole theory] may seem at first it is certainly
impossible to hold without introducing other profound modifications to square
our theory with the observed facts. Indeed, according to it the mass of the
proton should be the same as the mass of the electron; furthermore . . . this

hypothesis leads to the essential equivalence of positive and negative electricity under all circumstances – even on taking the interaction between matter and radiation rigorously into account . . . positive and negative electricity have essentially the same properties in the sense that the laws governing them are invariant under a certain substitution which interchanges the quantum numbers of the electrons with those of the protons. The dissimilarity of the two kinds of electricity thus seems to hide a secret of Nature . . . I fear that the clouds hanging over this part of the subject will roll together to form a new crisis in quantum physics.[30]

Dirac found the secret by cutting the knot: 'a hole, if there were one, would be a new kind of particle, unknown to experimental physics, having the same mass and opposite charge to an electron.'[31] It had been an unavoidable but not an easy road.

The discovery of the positron came two years later.[32] It marks the beginning of a pause, to last until 1947, in the interplay between fresh experimental impetus and the further development of the theory of the electron. The intervening years were a period of consolidation and at the same time of renewed speculation. On the one hand, important consequences of the positron theory were brought to light, as will be recalled in section 5. On the other, new attempts were developed to make the theory finite. These latter efforts sometimes sidestepped the concurrent developments in the positron theory, lending a curious diversity to the theoretical work in the 1930s. It is worth while to mention briefly at least some of these attempts, if it were only to show the evident divergences in opinion of that period in regard both to how things actually stood and to how to proceed next.

4

How to construct a finite relativistic electron theory – that is the common goal of three distinct attempts to be discussed next. The first one is based on a new formal definition of what one shall mean by going to the point model limit. It is of enduring interest. The second one is a proposal to modify the electromagnetic field Lagrangian in such a way that its customary form emerges only in the low frequency limit. Technical difficulties have made it impossible to give a satisfactory quantum formulation. The third one is the 1938 classical theory of Dirac. It is, at this late date, the first proposal for a rigorous and fully covariant classical equation of motion of a charged point particle, especially novel in that it needs not only initial but also final conditions to specify admissible solutions. Remarkable as this equation is in its own right, it has not led to further new physical insights.

(1) λ-limiting process. In 1933, Wentzel raised the interesting question: if one considers the point electron as the limit in which a finite electron radius tends to zero, then how do the answers depend on whether this radius is considered as a space like vector (as had been conventional) or as a time like vector?[33] He proceeded to show, initially in classical context, then in quantum theory, that the classical $1/a$ singularity is eliminated if electromagnetic field quantities on the world lines of an electron are defined by first going off the line in a time like distance, after which the limit is taken in which this distance goes to zero. The method is equivalent to an analytic continuation procedure due to Marcel Riesz.[34] The quantized version was brought into Hamiltonian form by Dirac.[35] However, new and typical quantum singularities now appear. It was proposed by Dirac to circumvent these by introducing an indefinite metric in Hilbert space.[36] Even so, there remained many difficulties as discussed by Pauli.[37] A more recent reconsideration of the indefinite metric method falls outside the plan of this review.[38]

(2) Studies were made by Born and co-workers of non-linear modifications of the Maxwell equations.[39] These were motivated by 'the conviction of the great philosophical superiority of the unitary idea'.[40] Noting the difficulties connected with the infinities and the unexplained facts of the existence of elementary particles and the structure of nuclei the view is expressed that 'the present theory (formulated by Dirac's wave equation) holds as long as the wave lengths . . . are long compared with the "radius of the electron" e^2/mc^2, but breaks down for a field containing shorter waves. The non-appearance of Planck's constant in this expression for the radius indicates that in the first place the electromagnetic laws are to be modified; the quantum laws may then be adapted to the new field equations'.[40] These ideas gave rise to a number of interesting classical studies;[41] but many were the difficulties encountered in the quantization.[42]

(3) In 1938 Dirac returned once more to the classical theory of the electron and commented that 'One of the most attractive ideas in the Lorentz model of the electron, the idea that all mass is of electromagnetic origin, appears to be wrong, for two separate reasons':[43] Why should the neutron's mass be electromagnetic? How could one hope to obtain the symmetry between positive and negative mass values as in the electron–positron theory? 'The departure from the electromagnetic theory of mass [Dirac continues] removes the main reason we have for believing in the finite size of the electron.'

What about the infinity difficulty of the point model?

One may think that this difficulty will be solved only by a better understanding of the structure of the electron according to quantum laws. However, it seems more reasonable to suppose that the electron is too simple a thing for the question of the laws governing its structure to arise, and thus quantum mechanics should not be needed for the solution of the difficulty – our easiest path of approach is to keep within the confines of the classical theory.

Dirac's 1938 program was not to modify the Maxwell–Lorentz equations, but rather, to seek a new interpretation for them. (The germs of these ideas go back to the work of Frenkel mentioned earlier.)[16] His method rests on the invariant separation of fields into proper and external parts. The latter is half the sum of the retarded and advanced interaction (as is also the case in some action at a distance models[44]). His final rigorous equation is

$$m\dot{v}_\mu - \tfrac{2}{3}e^2\ddot{v}_\mu - \tfrac{2}{3}e^2\dot{v}_\lambda{}^2 v_\mu = ev_\nu F_{\mu\nu},$$

where v_μ is the electron velocity, $F_{\mu\nu}$ the field incident on the electron. This equation looks the same as Lorentz' equation of motion, but actually it has different properties. In Lorentz' case terms proportional to positive powers of the electron radius a had to be neglected. There are no such terms here: a negative mechanical mass has been introduced so as to make the observed mass equal to m, after which the limit $a \to 0$ is taken. The occurrence of this sink of mechanical energy is closely connected with the appearance of 'run away solutions' of Dirac's *exact* equation of motion, i.e. solutions corresponding to accelerations even in the absence of external fields. This can readily be seen from the general integration of the above equation with $F_{\mu\nu}=0$. It is then that Dirac is obliged to impose a final condition on the acceleration, along with initial conditions on position and velocity, in order to single out the acceptable solutions $v_\mu = \text{constant}$.

Dirac's subsequent concern with the quantization of his theory led to his involvement with the λ-process and with the indefinite metric mentioned earlier.

5

Meanwhile it was becoming evident that the electromagnetic mass problem belongs entirely to the quantum domain and is altogether beyond the reach of classical correspondence arguments.

The study of the implications of the positron theory had proceeded apace. Vacuum energies and charge densities proved to be harmless quantities. A new singularity was diagnosed related to the polarization of the vacuum. New formal techniques were explored, such as what is now called the point-splitting method, first invented by Dirac,[45] and the use of effective non-linear Lagrangians. New effects were studied, such as the scattering of light by light and of light in a Coulomb field. And to order e^2 the electromagnetic mass was recalculated on the basis of the new theory. There exists a useful collection of reprints of the period.[46]

For the purpose of a final visit to the past, consider the expression (to order e^2) for the electromagnetic mass computed by the now old fashioned method of cutting of all virtual momenta at a value $\hbar/a$, where a symbolizes the electron radius. One finds:[47]

$$\delta mc^2 = \frac{3}{2\pi} \cdot \frac{e^2}{\hbar c} \cdot mc^2 \cdot 1n\left[\frac{\hbar}{mca} + \sqrt{\left(1 + \frac{\hbar^2}{m^2 c^2 a^2}\right)}\right]. \tag{3}$$

From the accuracy to which quantumelectrodynamics agrees with experiment it is now known that, whatever modifications the future may bring, it must be at an $a \ll \hbar/mc$. It is therefore entirely illegitimate to expand equation (3) under the assumption that

$$\frac{\hbar}{mca} \ll 1. \tag{4}$$

However, it is illuminating to see what happens formally under the assumption of the 'classical limit condition' (4):

$$\delta mc^2 \simeq \frac{3}{2\pi}\left[\frac{e^2}{a} - \frac{2}{3}\frac{\mu^2}{a^3} \cdots\right],$$

with $\mu = e\hbar/2mc$. Thus, as the result of an illegitimate expansion we recognize for the last time the term with which Lorentz set out; and also the magnetic term upon which Rasetti and Fermi reflected – but now it has turned negative. Legitimately, the problem of electromagnetic mass must forever reside in the domain of relativistic quantum theory; or beyond.

The m dependence of the right-hand side in equation (3) indicates how strongly quantum effects modify any unitary picture. To all orders in e:

$$m = mf\left(\frac{\hbar}{mca}, \frac{e^2}{\hbar c}\right).$$

It has been speculated that the function f might have a finite limit for $a \Rightarrow 0$:[48] and that, if so, the requirement that the entire mass of the electron should be electromagnetic would give a condition on the value of the fine structure constant.

6

Finally, already in the times under discussion, some theoretical ideas are encountered which were to come to full development only in the subsequent renormalization period.

(1) As early as 1930, Oppenheimer raised the question whether the radiative corrections to the difference in energy levels of an atom could be finite in spite of the fact that the proper energy of each level diverges.[49] This was prior to the development of the positron theory. Therefore, the question had to be answered in the negative.

(2) In 1937 Kramers 'attempted to represent the theory in such a way that the question of the structure and finite extension of the particles does not appear explicitly and that the quantity which is introduced as "particle mass" is from the very beginning the experimental mass'.[50]

I still can hear Kramers lament [wrote Uhlenbeck[51]] . . . he just could not understand why Dirac's radiation theory was so good, even though it made insufficient distinction between the proper field of the electron and the external field, even though it did not give an account for the infinite electromagnetic mass; and even though the correspondence with the classical theory of the electron was entirely unclear . . . He was aware earlier and more deeply than most of his contemporaries of the imperfections of the theory.

Kramer's 'structure-independent' approach is indeed the first version of the renormalization theory[52] – based however on the principle to cure the classical theory first, and to quantize later. Several problems were treated with the help of this new approach, notably the displacement of spectral lines.[53] Kramers was to come close to the real answers.[54]

(3) In 1944 it was noted by the writer that the electromagnetic self energy can be made finite to all orders in e^2 if the electromagnetic interaction is modified so as to include also a 'subtractive' short range vector field.[55] This was to become known as the regularization of the photon propagator. Some of the problematics of giving a realistic interpretation of such a theory were noted. This led to a further examination of compensation effects by other than vector fields and a first attempt to explain the proton–neutron mass difference.[56]

This concludes an account of the history of the first fifty years. As stated, the subsequent realization, stimulated by new experimental results, that quantum electrodynamics has a much larger domain of validity than was generally expected earlier is part of another story. So is the most recent revival of interest in finite theories. Suffice it to say that we are still unable to account for electromagnetic mass differences; and that, more generally a coherent theory of particle masses still eludes us. Perhaps, at some future time, the advantage of hindsight may make it seem stirring to recall, once in a while, struggles such as ours.

NOTES AND REFERENCES

1. G. G. Stokes, *Mathematical and Physical Papers*, vol. 1, p. 17 (reprinted by the Johnson Reprint Corp., New York and London: 1966).
2. J. J. Thomson, *Phil. Mag.* **11**, 227 (1881).
3. J. J. Thomson, *Phil. Mag.* **44**, 293 (1897). For the experimental history see, e.g. G. P. Thomson, 'The septuagenarian electron', *Phys. Today* (May 1967), p. 55.
4. H. A. Lorentz, *The Theory of Electrons*, p. 40 (reprinted by Dover Publications, New York: 1952).
5. M. Abraham, *Ann. Physik* **10**, 105 (1903). The specific numerical coefficients in equations (1) and (2) refer to a surface charge distribution. None of the qualitative statements depend on the kind of distribution.
6. See W. Kaufmann, *Ann. Physik* **19**, 487 (1906), also for literature on earlier experimental work.
7. See H. A. Lorentz, *Collected Works*, vol. v, p. 172; Martinus Nyhoff ed. (The Hague: 1937).
8. Bucherer and Langevin had studied an extended electron model with Lorentz contraction but with constant volume: A. H. Bucherer, *Mathematische Einführung in die Elektronentheorie*, pp. 57, 58 (Leipzig: 1904); P. Langevin, 'La physique des Electrons', *Rev. gen. Sci. pures et appl.* **16**, 257 (1905).
9. Ref. 4, p. 43. See further ref. 4, p. 339.
10. Ref. 4, p. 213.
11. A. Einstein, *Jahrbuch der Radioaktivität und Elektronik*, **4**, 411 (1907).
12. H. A. Lorentz, 'Het relativiteitsbeginsel', p. 93; Leyden lectures 1910–12 (E. J. Brill, Leyden: 1922).
13. H. Poincaré, *Rend. Palermo*, **21**, 29, 1906. The discussion of the Poincaré-stress presented here follows more closely the treatment of Lorentz, ref. 4, pp. 213–15. I want to thank Professor G. Holton for drawing my attention to Fermi's concern with this question. See, e.g. E. Fermi, *Collected Works*, vol. i, p. 24 (University of Chicago Press: 1962).
14. H. A. Lorentz, *Collected Works*, vol. v, p. 314; see also ref. 4, p. 335.
15. Ref. 4, p. 215.
16. J. Frenkel, *Z. Physik*, **32**, 518 (1925).
17. J. Frenkel, *Naturwiss.* **12**, 882 (1924).
18. See ref. 4, p. 217.
19. See, e.g. S. Goudsmit and G. E. Uhlenbeck, *Physica* **6**, 273 (1926).
20. M. Abraham, *Ann. Physik* **10**, 105 (1903); see G. E. Uhlenbeck and S. Goudsmit, *Naturwiss.* **13**, 953 (1925), footnote 3.
21. L. H. Thomas, *Nature* **107**, 514 (1926); *Phil. Mag.* **3**, 1 (1927).
22. F. Rasetti and E. Fermi, *Nuovo Cimento* **3**, 226 (1926); E. Fermi, *Collected Works*, vol. i, p. 212 (University of Chicago Press: 1962).
23. H. A. Lorentz, *Collected Works*, vol. vii, p. 179.

24. P. A. M. Dirac, *Proc. Roy. Soc. (London)* A117, 610 (1928); A118, 351 (1928).
25. H. Weyl, *Z. Physik* 56, 332 (1929).
26. P. A. M. Dirac, *Proc. Roy. Soc. (London)* A126, 360 (1929).
27. P. A. M. Dirac, *Proc. Cambridge Phil. Soc.* 26, 361 (1930).
28. I. Tamm, *Z. Physik* 62, 545 (1930).
29. J. R. Oppenheimer, *Phys. Rev.* 35, 562 (1930).
30. H. Weyl, *The Theory of Groups and Quantum Mechanics*, 2nd ed. pp. 263–4 and preface (Dover Publ. Inc. New York).
31. P. A. M. Dirac, *Proc. Roy. Soc. (London)* A133, 60 (1931).
32. C. D. Anderson, *Science* 76, 238 (1932); *Phys. Rev.* 43, 491 (1933).
33. G. Wentzel, *Z. Physik* 86, 479, 635 (1933); 87, 726 (1934).
34. This equivalence was shown by S. T. Ma, *Phys. Rev.* 71, 787 (1947). This paper contains references to and an exposé of Riesz' method. See further T. Gustafson, *Kgl. Fysisk Sallsk. Lund. Forh.* 16, no. 2 (1946).
35. P. A. M. Dirac, *Ann. Inst. H. Poincaré* 9, 13 (1939); 'Quantum electrodynamics', *Comm. Dublin Inst. for Adv. Studies*, A1 (1943).
36. P. A. M. Dirac, *Proc. Roy. Soc. (London)* A180, 1 (1942).
37. W. Pauli, *Rev. Mod. Phys.* 15, 175 (1943). See further P. A. M. Dirac, *Comm. Dublin Inst. for Adv. Studies*, A3 (1946).
38. T. D. Lee and G. C. Wick, *Phys. Rev.* D2, 1033 (1970).
39. For a review and further literature see M. Born, *Ann. Inst. H. Poincaré*, 7, 155 (1937).
40. M. Born and L. Infeld, *Proc. Roy. Soc. (London)* A144, 425 (1934).
41. See further E. Schroedinger, *Proc. Roy. Soc. (London)* A150, 465 (1935); L. Infeld, *Proc. Cambridge Phil. Soc.* 32, 127 (1936); 33, 70, (1937); M. H. L. Pryce, *Proc. Cambridge Phil. Soc.* 31, 50, 625 (1935).
42. See e.g. M. H. L. Pryce, *Proc. Roy. Soc. (London)* A159, 355 (1937).
43. P. A. M. Dirac, *Proc. Roy. Soc. (London)* A167, 148 (1938).
44. J. A. Wheeler and R. P. Feynman, *Rev. Mod. Phys.* 17, 157 (1945).
45. P. A. M. Dirac, Rapports du 7ième Conseil Solvay, 1933; *Proc. Cambridge Phil. Soc.* 30, 150 (1934).
46. For all of these topics see *Selected Papers on Quantum Electrodynamics*, J. Schwinger ed. (Dover Publ. New York: 1958).
47. V. Weisskopf, *Z. Physik* 89, 27 (1934); 90, 817 (1934); *Phys. Rev.* 56, 72 (1939).
48. G. Racah, *Phys. Rev.* 70, 406 (1946).
49. J. R. Oppenheimer, *Phys. Rev.* 35, 461 (1930).
50. H. A. Kramers, *Nuovo Cimento* 15, 108 (1938); *Collected Scientific Papers*, p. 831 (North-Holland Publishing Co.: 1956).
51. G. E. Uhlenbeck, *Oude en Nieuwe Vragen der Natuurkunde* (North-Holland Publishing Co.: 1955).
52. H. A. Kramers, *Quantum Mechanics* §§89, 90 (North-Holland Publishing Co.: 1957).
53. J. Serpe, *Physica* 7, 133 (1940); 8, 161 (1941); W. Opechowski *Physica* 8, 226 (1941).
54. H. A. Kramers, *Collected Scientific Papers*, p. 858.
55. A. Pais, *Phys. Rev.* 68, 227 (1945). The detailed discussion is given in A. Pais, 'On the theory of elementary particles', *Trans. Roy. Acad. Sciences of the Netherlands*, 19, no. 1 (1947).
56. See ref. 55 and S. Sakata and O. Hara, *Progr. Theoret. Phys. (Kyoto)* 2, 30 (1947).

6

The Dirac Equation

A. S. Wightman

The purpose of this paper is twofold. On the one hand, following a suggestion of Professor Abdus Salam, I give an answer to the question: Why was the Dirac equation such a great discovery? However presumptuous this may appear, it is a pleasure to do it in celebration of Professor Dirac's seventieth birthday. The second objective of the paper is to add a few morsels to our present knowledge of the description of leptons.

1. AN APPRECIATION OF THE DIRAC EQUATION

Let me begin with a retrospective view of the Dirac equation, by answering the historical question: Why was the discovery of the Dirac equation[1,2] so important for the physics of the 1920s and 30s? The answer is manifold. The Dirac equation provided a relativistic description of spin $\frac{1}{2}$ particles and in particular of the electron. In so doing, it gave a relativistic description of spin and opened the way for the application of group theory to the description of particles of arbitrary spin. The reinterpretation of the Dirac equation as a field equation that followed from Dirac's theory of holes was decisive in the conceptual transformation of single particle theory to many particle (quantum field) theory. The resulting quantum electrodynamics of spin $\frac{1}{2}$ particles, refined by two generations of theoretical work, is the best theory we have. Although it is an approximation since it does not include the effects of weak and strong interactions it has survived many stringent experimental tests when applied to electrons and muons. Let me expand on these statements.

THE RELATIVISTIC DESCRIPTION OF ELEMENTARY PARTICLES
Consider first the Dirac equation in the absence of interaction

$$[-i\gamma^\mu \partial_\mu + m]\psi(x) = 0 \qquad (1.1)$$

with the scalar product[3]

$$(\psi, \chi) = \int_\Sigma d\Sigma^\mu(x)\psi^\dagger(x)\gamma_\mu\chi(x). \qquad (1.2)$$

Its positive energy solutions, ψ, such that $(\psi, \psi) < \infty$ provide a description of a free spin $\frac{1}{2}$ particle consistent with the requirements of the special theory of relativity. The spin appears as a consequence of the transformation law of the solutions under rotations, which in turn is determined by the transformation law under homogeneous Lorentz transformations. Expressing this idea in the language of the twenties one could say that the spin of the electron revealed itself as a consequence of the relativistic invariance of the Dirac equation. The only relativistic equation for a massive particle that preceded the Dirac equation, the Klein–Gordon equation

$$(\square + m^2)\phi(x) = 0 \tag{1.3}$$

for a complex valued function ϕ, describes a spinless particle so this insight was impossible for it.

Nowadays, group theory having permeated theoretical physics and the representation theory of the Poincaré group being well known, one might describe the situation in terms of the Poincaré group, $\mathscr{P}_+^\uparrow$, or more accurately, its universal covering group, the inhomogeneous complex unimodular group in two dimensions, ISL(2, $\mathbb{C}$). The action of the transformation $\{a, A\}$ of this group on the positive energy solutions of the Dirac equation $\psi \rightarrow U(a, A)\psi$

$$(U(a, A)\psi)(x) = S(A)\psi(\Lambda(A^{-1})(x - a)) \tag{1.4}$$

defines a continuous irreducible unitary representation of ISL(2, $\mathbb{C}$). If such representations have a mass greater than zero, they are uniquely determined up to unitary equivalence by a spin, j, where $j = 0, \frac{1}{2}, 1, \frac{3}{2}, \dots$ Representations of different mass or spin are inequivalent. Thus the spin along with the mass turns out to be a group invariant characterizing the unitary representation of the relativity group associated with the wave equation. Many years passed and a now well known analysis of the representations of ISL(2, $\mathbb{C}$) was necessary before this statement could be made in such generality[4] but it is very clear that the starting point was the Dirac equation.

FROM SINGLE PARTICLE EQUATIONS TO QUANTIZED FIELDS

The historical developments that led from the Dirac equation as a description of a single particle to the Dirac equation as an equation for coupled quantized fields surely constitute one of the most fascinating conceptual transformations in the history of theoretical physics. As a first step in describing it, let me recall an argument that Dirac gave for his equation. He said (with a slight change of notation)

The general interpretation of non-relativity quantum mechanics is based on the transformation theory, and is made possible by the wave equation being of the form

$$\left(H - i\frac{\partial}{\partial x^0}\right)\psi = 0 \qquad (1.5)$$

i.e. being linear in $\partial/\partial x^0$, so that the wave function at one time determines the wave function at any later time. The wave equation of the relativity theory must also be linear in $\partial/\partial x^0$ if the general interpretation is to be possible.[5]

At first sight, this argument would appear to exclude the Klein–Gordon equation (1.3). Since it is of second order, apparently one has to give both ϕ and $\partial\phi/\partial x^0$ at time $t=0$ to determine a solution later. That is true but not really to the point since the positive energy solutions of (1.3) all satisfy

$$\left[\sqrt{(-\Delta + m^2)} - i\frac{\partial}{\partial x^0}\right]\phi = 0, \qquad (1.6)$$

and therefore $\partial\phi/\partial t$ is determined at time t if ϕ is given at that time.

Of course, the set of positive energy solutions of the Klein–Gordon equation is not complete as a single particle theory until one gives the scalar product. Now one writes

$$(\phi, \chi) = \frac{i}{2}\int_\Sigma d\Sigma^\mu(x)[\overline{\phi}(x)\partial_\mu\chi(x) - \partial_\mu\overline{\phi}(x)\chi(x)] \qquad (1.7)$$

and regards the resulting theory as being a consistent description of a free spin zero particle, just as consistent in its own way as the Dirac theory is for a particle of spin $\frac{1}{2}$. The essential point is that $(\phi, \phi) > 0$ for ϕ a positive energy solution, even though $(\phi, \phi) < 0$ for ϕ a negative energy solution.

At the time Dirac wrote his paper the Klein–Gordon current had been used in the correspondence theory of radiation but no probability interpretation had been developed.[6, 7] It was natural therefore that Dirac should regard his theory as more satisfactory in this respect. Whereas the fourth component of the Klein–Gordon current,

$$\frac{i}{2}(\overline{\phi}\partial_0\phi - \partial_0\overline{\phi}\phi), \qquad (1.8)$$

is not positive definite, the analogous expression $\psi^\dagger\gamma^0\psi$ in Dirac theory is. In fact, if γ^0 is chosen as hermitean and γ^j, $j=1, 2, 3$ as anti-hermitean, $\eta = \gamma^0$ and

$$\psi^\dagger(x)\gamma^0\psi(x) = \Sigma_{\alpha=1}^4 |\psi_\alpha(x)|^2, \qquad (1.9)$$

which is clearly analogous to the expression $|\psi(x)|^2$ for the probability

H

density of non-relativistic Schrödinger theory. It is implicit in this discussion that (1.9) is the probability density to find an electron at $\mathbf{x}$ at time x^0. Long after it was realized that for single particle theory in which only positive energy wave functions are used,[8] this assumption cannot be correct, because multiplication by $\mathbf{x}$ does not carry positive energy wave functions into positive energy wave functions. To describe the localization of particles satisfying relativistic wave equations it is necessary to introduce quite a different coordinate operator. In this respect the Klein–Gordon equation is neither better nor worse than the Dirac equation.

The case of a free particle is only a first step toward a theory of interaction. A second step is the theory with a time-independent external field, a model appropriate to the description of a particle moving in the electric field of a nucleus. Here the Dirac theory had a famous success since it gave the Sommerfeld formula for the bound states of hydrogen-like atoms.[9] For sufficiently weak external potentials it was possible to separate the positive and negative energy solutions and, restricting attention to the positive energy ones, to have a consistent single particle theory. (For the Coulomb potential, $A^0 = eZ/r$, the restriction turned out to be

$$Z < \frac{\sqrt{3}}{2} \frac{\hbar c}{e} \approx \frac{\sqrt{3}}{2} \, 137.)$$

The Sommerfeld formula was in good agreement with the then existing experiments and in fact was not improved upon until the discovery of the Lamb shift and its explanation in terms of radiative corrections in quantum electrodynamics. For the Klein–Gordon equation, as far as internal consistency was concerned, the situation was again neither better nor worse. For an electrostatic potential the equation is

$$[(\partial^0 - eiA^0)^2 - \Delta + m^2]\phi = 0, \tag{1.10}$$

and so, if one introduces $\phi(x) = u(\mathbf{x}) \exp -(iEt)$, u satisfies

$$[-(E + eA^0)^2 - \Delta + m^2]u = 0, \tag{1.11}$$

and the scalar product between two such solutions is

$$(\phi_1, \phi_2) = \int d^3x u_1(\mathbf{x})[((E_1 + E_2)/2) + eA^0]u_2(\mathbf{x}). \tag{1.12}$$

This is not an eigenvalue problem of the usual kind since the 'eigenvalue' E appears quadratically in (1.11) and hence shows up in the normalization condition (1.12). Nevertheless, for sufficiently weak eA^0, the spectrum of Es is real and the scalar product is positive on the subspace spanned by the positive energy solutions.[10]

Viewed in hindsight, the arguments for the superiority of the Dirac equation over the Klein–Gordon equation interpreted as single particle theories are not convincing. Of course, the Dirac equation had the obvious advantage that it described the electron spin, while at the time there was no known candidate for a spin zero particle.

However successful the Dirac equation may be when it is interpreted as a single particle theory for weak time-independent external fields, it encounters trouble, in general, for time-dependent external fields. The difficulty is what now goes under the general label of the Klein Paradox: Time-dependent external fields cause transitions from positive to negative energy states.[11] There is no consistent way to amputate the negative energy states of the theory so as to prevent such transitions. This is a difficulty both for the Dirac equation and for the Klein–Gordon equation, but it was Dirac's work on his own theory which showed the way for particle theories in general.

Dirac's way out was hole theory.[12] In it the negative energy states are thought of as all occupied so transitions of positive energy electrons into negative energy states are forbidden by the Pauli Exclusion Principle. Why do we not see the vast sea of negative energy particles? Because it is declared invisible! With this interpretation a hole in the negative energy sea is seen as a particle of opposite charge and is produced when an external perturbation causes a negative energy particle to make a transition to a positive energy state, the net result being a pair consisting of a positive energy particle and a hole.

It is difficult for one who, like me, learned quantum electrodynamics in the mid 1940s to assess fairly the impact of Dirac's proposal. I have the impression that many in the profession were thunderstruck at the audacity of his ideas. This impression was received partly from listening to the old-timers talking about quantum-electrodynamics a decade and-a-half after the creation of hole theory; they still seemed shell-shocked. Pauli's considered opinion (before the discovery of the positron!) was expressed in his article on wave mechanics:[13]

. . . Neuerdings versuchte Dirac . . . den bereits von Oppenheimer diskutierten Ausweg, die Löcher mit antielektronen, Teilchen der Ladung $+e$ und der Elektronenmasse, zu identifizieren. Ebenso musste es dann neben den Protonen noch Antiprotonen geben. Das tatsächliche Fehlen solcher Teilchen wird dann auf einen speziellen Anfangszustand zurückgeführt, bei dem eben nur die eine Teilchensorte vorhanden ist. Dies erscheint schon des halb unbefriedigend, weil die Naturgesetze in dieser Theorie in bezug auf Elektronen und Antielektronen exakt symmetrisch sind. Sodann müssten jedoch (um die Erhaltungssätze von Energie und Impuls zu befriedigen mindestens zwei) γ-Strahl-Photonen sich von selbst in ein Elektron und

Antielektron umsetzen können. Wir glauben also nicht, dass dieser Answeg ernstlich in Betracht gezogen werden kann.

After the discovery of the positron, it became clear that serious work on the quantum electrodynamics of spin $\frac{1}{2}$ particles had to be based on hole theory or something very like it. The ground was well prepared. Jordan had been advocating the introduction of field operators for the electrons for some time,[14] and the work of Jordan and Wigner[15] had provided the key to the description of particles satisfying Fermi–Dirac statistics: the anti-commutation relations. The way was open to write the theory of electrons and positrons in a form completely symmetric under the exchange of particle and antiparticle, a form in which the infinite sea of negative energy electrons has vanished from the theory except as a poetic description of the prescription for forming the electromagnetic current. At the risk of being pedantic let me describe the resulting formulae. The Dirac field operator is

$$\psi(x) = (2\pi)^{-3/2} \int d\Omega_m(p)$$

$$\sum_\zeta [a(p, \zeta)u(p, \zeta) \exp{-ip \cdot x} + b^*(p, \zeta)u^c(p, \zeta) \exp{ip \cdot x}], \qquad (1.13)$$

where $d\Omega_m(p) = [\mathbf{p}^2 + m^2]^{-\frac{1}{2}}d^3p$ and the integral goes over the positive energy mass shell $p^0 = [\mathbf{p}^2 + m^2]^{\frac{1}{2}}$. The $u(p, \zeta)$, $\zeta = \pm 1$ are two linearly independent solutions of the momentum space Dirac equation for momentum p.

$$(-\not{p} + m)u(p, \zeta) = 0, \quad \not{p} = \gamma^\mu p_\mu, \qquad (1.14)$$

$u^c(p, \zeta)$ is the charge conjugate of $u(p, \zeta)$ defined by

$$u^c(p, \zeta) = C^{-1}u(p, \zeta), \qquad (1.15)$$

where C is the charge conjugation matrix defined by

$$-\bar\gamma^\mu = C\gamma^\mu C^{-1}, \qquad (1.16)$$

$a(p, \zeta)$ is the annihilation operator for an electron in the state described by $u(p, \zeta)$; $b^*(p, \zeta)$ is the creation operator for an electron in that same state. The annihilation and creation operators satisfy the anti-commutation relations

$$[a(p, \zeta), a(q, \rho)]_+ = 0 = [b(p, \zeta), b(q, \rho)]_+ = [a(p, \zeta), b(q, \rho)]_+ \qquad (1.17)$$

$$[a(p, \zeta), a^*(q, \rho)]_+ = p^0\delta(\mathbf{p} - \mathbf{q})\delta_{\rho\zeta} = [b(p, \zeta), b^*(q, \rho)]. \qquad (1.18)$$

The vacuum state, Ψ, of the theory has no positrons and electrons and satisfies

$$a(p, \zeta)\Psi_0 = 0$$
$$b(p, \zeta)\Psi_0 = 0. \qquad (1.19)$$

As a consequence of the commutation relations (1.17) and (1.18) for the annihilation and creation operators, the field $\psi(x)$ itself satisfies the anti-commutation relations

$$[\psi(x), \psi(y)]_+ = 0 \quad [\psi(x), \psi^\dagger(y)]_+ = (1/i)S(x-y), \tag{1.20}$$

where

$$S(x) = (-\gamma^\mu \partial_\mu + m)\Delta(x) \tag{1.21}$$

and

$$\Delta(x) = \Delta^{(\dagger)}(x) + \Delta^{(-)}(x) \tag{1.22}$$

with

$$\Delta^{(\pm)}(x) = \frac{(\pm i)}{2(2\pi)^2} \int d\Omega_m(p) \exp -ip \cdot x. \tag{1.23}$$

The transformation law of the field under ISL(2, $\mathbb{C}$) is

$$U(a, A)\psi(x)U(a, A)^{-1} = S(A^{-1})\psi(\Lambda(A)x + a), \tag{1.24}$$

where $\{a, A\} \to U(a, A)$ is the transformation law of the states of the theory. The charge symmetry is expressed by a unitary operator U_c which satisfies

$$U_c a(p, \zeta) U_c^{-1} = b(p, \zeta) \tag{1.25}$$

and commutes with $U(a, A)$, and leaves the vacuum invariant

$$U_c \Psi_0 = \Psi_0. \tag{1.26}$$

U_c evidently interchanges the positrons and electrons. Formulae essentially equivalent to these were written by Fock in 1933.[16]

The essential change in the formula for ψ as compared with the pre-hole theory expression is the appearance in (1.13) of the positron creation operators $b^*(p, \zeta)$ in the second set of terms; in the pre-hole theory expressions there would have been an annihilation operator for negative energy electrons. The prescription for the electric current referred to above is

$$j^\mu(x) = :\psi^\dagger \gamma^\mu \psi:(x) = \lim_{y \to x} [\psi^\dagger(x)\gamma^\mu\psi(y) - (\Psi_0, \psi^\dagger(x)\gamma^\mu\psi(y)\Psi_0)], \tag{1.27}$$

the subtracted term was thought of as the contribution from the negative energy sea. Today we have become reconciled to the fact that quantities like $\psi^\dagger(x)\gamma^\mu\psi(y)$ have singularities on the diagonal $x = y$, whose removal via a prescription such as (1.27) may leave a well-defined and physically meaningful operator. To reassure the skeptical student that all this makes mathematical sense one simply realizes all the indicated operators in Fock space and proves it directly.[17]

The new way of looking at the Dirac equation did not arouse the widespread satisfaction one might have expected. Dirac himself was inclined to work directly with the infinite sea[18] and resisted the idea of the Heisenberg and Pauli paper[19] that quantum electrodynamics should be regarded as the dynamics of coupled quantized fields.[20] Pauli was dissatisfied with the prescriptions of hole theory, regarding them as ugly and artificial. He coined the derisory phrase 'subtraction physics' to describe hole theory. With Weisskopf he developed a quantum field theory of a scalar field describing charged spin zero particles.[21] It, like the hole theory, predicted such phenomena as pair production by an external field. For reasons that now appear a bit incomprehensible since, as Pauli himself showed, the theories can be constructed in parallel, he regarded the treatment of subtractions in the Pauli–Weisskopf theory as natural, and he dubbed it the *anti-Dirac* theory.[22] Nevertheless, it was clear that a revolution had taken place. To describe the electron with full precision one used the Dirac equation, but the Dirac equation for a field. Heisenberg adopted the point of view of the charge symmetrical theory whole-heartedly in his profound study of quantum electrodynamics.[23]

RENORMALIZATION THEORY

Some of the basic ideas of renormalization theory go back to the nineteenth century. For example, the notion that the interaction between an object and its surroundings can cause an alteration in the object's inertia was discussed for electrically charged bodies by J. J. Thomson in 1881.[24] Much effort was devoted to the idea in the context of the Abraham and Lorentz models of the electron.[25] However, it was Dirac who first found the phenomenon of vacuum polarization and showed that it gives rise to a renormalization of charge.[26] Dirac's investigation of this problem was carried out for a Dirac electron interacting with an external electromagnetic field. The field ψ then satisfies

$$[-i\gamma^\mu(\partial_\mu - eiA_\mu(x)) + m]\psi(x) = 0 \qquad (1.28)$$

instead of (1.1), where A_μ is the vector potential describing the external electromagnetic field.

For present and later purposes, one may as well consider a general perturbation given by a 4×4 matrix $B(x)$ whose entries are complex valued functions on space time:

$$[-i\gamma^\mu\partial_\mu + m + B(x)]\psi(x) = 0. \qquad (1.29)$$

For simplicity, it will be assumed that $B(x)$ is smooth (has derivatives of all orders) and vanishes rapidly at infinity in all directions in space time. Furthermore, $B(x)$ is required to satisfy

$$B(x)^* = \eta B(x) \eta^{-1}. \tag{1.30}$$

This guarantees the conservation of current in the theory. That turns out to be necessary in order that probability be conserved.

To bring out the significance of Dirac's results I describe the solution of this problem in somewhat different terms from those originally used. The first step is to replace (1.29) by an integral equation, the so-called Yang–Feldman equation.[27]

$$\psi(x) = \psi^{in}(x) - \int S_R(x-y)B(y)\psi(y)d^4y. \tag{1.31}$$

Here ψ^{in} is the free field to which ψ reduces for times in the distant past. S_R is

$$S_R(x) = (i\gamma^\mu \partial_\mu + m)\Delta_R(x), \tag{1.32}$$

$$\Delta_R(x) = \frac{1}{(2\pi)^4} \int [-p^2 + m^2]^{-1} \exp -ip \cdot x \, d^4p, \tag{1.33}$$

where the integral over p is taken as a retarded boundary value, i.e. it is the limit as $\Sigma \to 0$ of the expression with p^2 replaced by $(p^0 - i\epsilon)^2 - \mathbf{p}^2$. If ψ^{in} is regarded as given, say by the formulae (1.13) . . . (1.19), then the problem of solving (1.29) for the operator ψ can be reduced to solving a related equation for complex valued functions. To state this result a few definitions are necessary.

A *fundamental solution* of (1.29) is by definition a 4×4 matrix of distributions $E(x, y)$ depending on two space time variables and satisfying

$$[-i\gamma^\mu \partial_{x_\mu} + m + B(x)]E(x, y) = \delta(x-y)1 \tag{1.34}$$

and

$$[+i\partial_{y_\mu}E(x, y)\gamma^\mu + E(x, y)(m + B(y))] = \delta(x-y)1. \tag{1.35}$$

A *retarded* fundamental solution $S_R(x, y; B)$ is a fundamental solution that satisfies

$$S_R(x, y; B) = 0 \text{ for } (x-y)^2 < 0 \tag{1.36}$$
$$\text{and for } (x-y)^0 < 0.$$

Similarly an *advanced* fundamental solution $S_A(x, y; B)$ is one satisfying

$$S_A(x, y; B) = 0 \text{ for } (x-y)^2 < 0 \tag{1.37}$$

$$\text{and for } (x-y)^0 > 0.$$

A *weakly retarded* fundamental solution is a fundamental solution that is rapidly decreasing as $x-y$ approaches infinity in every direction outside the future light cone. That is $S_R(x, y; B)$ is weakly retarded if for each pair of test functions f, g each positive integer n and each vector ℓ outside the future light cone there exists a constant c such that

$$\left| \int f(x + \tau \ell) S_R(x, y; B) g(y) d^4x d^4y \right| < C/(1 + \tau^n) \tag{1.38}$$

for all $\tau > 0$. *Weakly advanced* is defined in the same way but with the future light cone replaced by the past light cone.

Now smear the equation (1.31) with the test function f and rewrite it as

$$\psi(T_R f) = \psi^{in}(f) \tag{1.39}$$

where

$$(T_R f)(x) = f(x) + \int d^4y\, f(y) S_R(y - x) B(x). \tag{1.40}$$

This T_R is a continuous linear map of the test function space into itself. If it has a continuous inverse then, the field ψ can be defined by

$$\psi(f) = \psi^{in}(T_R^{-1} f). \tag{1.41}$$

It can be shown that a necessary and sufficient condition that T_R^{-1} exist and be continuous is that the Dirac equation with the external field, B, possess a weakly retarded fundamental solution $S_R(x, y; B)$.[28] There is a simple formula for T_R^{-1} in terms of $S_R(x, y; B)$,

$$(T_R^{-1} f)(x) = f(x) - \int f(y) d^4y\, S_R(y, x; B) B(x). \tag{1.42}$$

If one works out the expression for the adjoint field $\psi^\dagger(f)$ it turns out to be

$$\psi^\dagger(f) = \psi^{in\dagger}(T_R'^{-1} f), \tag{1.43}$$

where

$$(T_R' f)(x) = f(x) + B(x) \int S_A(x - y) f(y) d^4y \tag{1.44}$$

and

$$(T_R'^{-1} f)(x) = f(x) - B(x) \int S_A(x, y; B) f(y) d^4y. \tag{1.45}$$

Since, as will be discussed later, unique weakly retarded and advanced fundamental solutions exist for the Dirac equation and are, in fact, retarded and advanced respectively, there are no infinities or ambiguities in the solution for the field ψ. Where then are the celebrated vacuum polarization infinities? They are in the formula for the current. To study them we look first at

$$\int \psi^\dagger(x) F(x, y) \psi(y) d^4x d^4y. \tag{1.46}$$

It is equal to

$$\int \psi^{in\dagger}(x) [(T_R'^{-1} \otimes T_R^{-1}) F](x, y) \psi^{in}(y) d^4x d^4y \tag{1.47}$$

by virtue of the expressions (1.41) and (1.43) for ψ and $\psi^\dagger$ in terms of

ψ^{in} and ψ^{int}. If the expressions (1.45) and (1.42) for $T_R'^{-1}$ and T_R^{-1} are inserted, (1.47) becomes

$$\int \psi^{int}(x)K(x, y)\psi^{in}(y)d^4x d^4y, \qquad (1.48)$$

where

$$K(x, y) = F(x, y) - B(x)\int S_A(x, z; B)d^4zF(z, y)$$
$$- \int d^4zF(x, z)S_R(z, y; B)B(y) \qquad (1.49)$$
$$+ B(x)\int\int S_A(x, z; B)d^4zF(z, w)d^4wS_R(w, y; B)B(y).$$

To obtain a candidate for the current one has to pass to the diagonal in the variables x, y i.e. to consider the limiting behavior for a sequence of test functions F_n satisfying $F_n(x, y) \to f(x)\delta(x-y)\gamma^\mu$. Formally, the result is

$$\int d^4xf(x)[\psi^{int}(x)\gamma^\mu\psi^{in}(x) - \int d^4y\psi^{int}(y)B(y)S_A(y, x; B)\gamma^\mu\psi^{in}(x)$$
$$- \int d^4y\psi^{int}(x)\gamma^\mu S_R(x, y; B)B(y)\psi^{in}(y) \qquad (1.50)$$
$$+ \int\int d^4y\psi^{int}(y)B(y)S_R(y, x; B)\gamma^\mu S_A(x, z; B)B(z)\psi^{in}(z)d^4z].$$

It was a slight modification of this expression evaluated in perturbation theory up to terms linear in B, that Dirac studied. The prescriptions of hole theory in the absence of an external field suggest that the first term be replaced by (1.27) i.e. that its expectation value in the vacuum, Ψ_0^{in}, of ψ^{in} be subtracted. Dirac showed that the approximate expression he was using for the second and third terms in (1.50) could be split in two, a part (infinite l) interpretable as a renormalization of charge and a second (finite) part. Later on the second part was shown by Uehling to give rise to small deviations from Coulomb's law for the force between two charges, an effect which has been detected in the Lamb shift.[29] The general prescription is obtained as follows. Write

$$\psi^{int}(x)\psi^{in}(y) = :\psi^{int}(x)\psi^{in}(y): + (\Psi_0^{in}, \psi^{int}(x)\psi^{in}(y)\Psi_0^{in}). \qquad (1.51)$$

Then the contribution to (1.50) from the terms with double dots is well defined, so it remains to discuss the vacuum expectation values. This requires a detailed study of the singularities possessed by the advanced and retarded fundamental solutions and of their contribution to the vacuum expectation value of (1.50). Suffice it to say that Dirac's calculation indicates the general situation: a perturbation theory calculation to a few low orders yields the only singular terms. Once they are subtracted the rest yields well defined contributions to the current.[30]

This sketchy account of the construction of the current in the external field problem will have to suffice to indicate the nature of renormalization, and the role that the Dirac equation played in the first stages of the theory. Of course, for coupled fields things are much more complicated and despite prodigious efforts over three decades many of the main questions are still open. It is striking, however, how the quantum electrodynamics of spin $\frac{1}{2}$ particles described by the Dirac equation has risen in the world. At the time of its original construction, the theory was regarded as surely internally inconsistent, and dramatic disagreement with experiment was expected for say Compton scattering of photons at $(\hbar c/e^2)mc^2 \approx 70\mathrm{MeV}$. No such catastrophes occurred in the cosmic ray experiments of the thirties and the cross-sections for the Compton effect for Bremsstrahlung and pair production were found to be in general agreement with experiment. When experiments became accurate enough to test radiative corrections the theory was refined to produce unambiguous predictions and they agreed. Things have now reached the stage at which the next generation of experiments will require knowledge of strong interaction theory for their description. Quantum electrodynamics of electrons and muons has turned out to be a very fine theory indeed.

To the preceding retrospective account of the importance of the Dirac equation, I want to add some remarks on properties which distinguish it among relativistic wave equations. One elementary kinematical fact comes immediately to mind. The Dirac field is well suited to the construction of other fields by taking tensor products since all kinds of tensors and spinors can be constructed in this way. A second fact is probably deeper. An equation for a field restricts the possible forms of relativistic coupling to other fields. Certain of these interactions may have distinguished regularity properties as well as other physical properties. For example, minimal coupling to a vector potential, A_μ, as in equation (1.28) yields a particle with gyromagnetic ratio $g = 2$, neglecting radiative corrections. When A_μ is quantized the theory is renormalizable. On the other hand, if one were willing to include a Pauli term $(g-2)(e/4m)\sigma_{\mu\nu}F^{\mu\nu}$ in (1.28), one could get an arbitrary real value for g but the theory would be non-renormalizable for $g \neq 2$. This favorable situation for $g = 2$ does not hold for all spin $\frac{1}{2}$ equations. For example, there is an irreducible wave equation with sixteen components for which minimal coupling yields a gyromagnetic ratio $g = \frac{3}{2}$.[31] A still more striking feature of the Dirac equation is its stability under smooth local perturbations, the subject of the last subsection of this section.

STABILITY UNDER SMOOTH LOCAL PERTURBATIONS

Consider again the Dirac equation (1.29), where $B(x)$ is a 4×4 matrix whose entries are smooth and rapidly decreasing (for brevity, we write

$B \in \mathscr{S}$ adapting a notation of L. Schwartz.[32] Such a $B(x)$ will be called a smooth local perturbation. By expanding B in the sixteen products of the γ^{μ} we get

$$B(x) = \rho(x)1 + \sigma(x)\gamma^5 - eA_\mu(x)\gamma^\mu + B_\mu\gamma^\mu\gamma^5 + \tfrac{1}{2}(g-2)\frac{e}{2m}F^{\mu\nu}(x)\sigma_{\mu\nu}. \quad (1.52)$$

The five uniquely determined coefficients are also in $\mathscr{S}$ and can be thought of as external fields with various transformation laws under $\mathscr{P}_+^\uparrow$. The stability of the Dirac equation under smooth local perturbation is expressed precisely in the following theorem.

Theorem. There exist unique retarded and advanced fundamental solutions of the Dirac equation (1.29) for every $B \in \mathscr{S}$. These fundamental solutions are tempered distributions and the mapping

$$\{f, B\} \to \int_A S_R(x, y; B)f(y)d^4y \quad (1.53)$$

of $\mathscr{S} \times \mathscr{S}$ into $\mathcal{O}_M$ is continuous. Here $\mathcal{O}_M$ is the set of infinitely differentiable functions all of whose derivatives are bounded in absolute value by polynomials.

Sketch of Proof. Consider the identity

$$[-i\gamma^\mu\partial_\mu + B(x) + m][-(-i\gamma^\mu\partial_\mu + B(x)) + m]$$
$$= m^2 - (-i\gamma^\mu\partial_\mu + B(x))^2 \quad (1.54)$$
$$= (\square + m^2)1 + \{i[\gamma^\mu, B(x)]_+\partial_\mu + i\gamma^\mu(\partial_\mu B)(x) - B(x)^2\}.$$

The last expression is the wave operator perturbed by a first order matrix operator. Suppose that it has a right fundamental solution $E(x, y)$. Then

$$[-i\gamma^\mu\partial_\mu + m + B(x)][i\gamma^\mu\partial_\mu + m - B(x)]E(x, y) = \delta(x - y)1, \quad (1.55)$$

so $[i\gamma^\mu\partial_\mu + m - B(x)]E(x, y)$ is a right fundamental solution for the Dirac equation (1.29) itself. Now for the wave operator $(\square + m^2)$ acting on a single complex-valued function, there is a standard argument which shows that the addition of a first-order perturbation $a_\mu(x)\partial^\mu + b(x)$ with smooth coefficients does not change the characteristics nor the singularities of the fundamental solution. Like the wave operator $\square + m^2$ itself, the operator $\square + m^2 + a^\mu\partial_\mu + b$ has a unique weakly retarded fundamental solution that is retarded. Furthermore the operator maps the set of C^∞ functions in an open set onto themselves and this implies

that a right fundamental solution is automatically a left fundamental
solution. (The essential ideas of the arguments are contained in Hörmander, chapters VII and VIII.[33]) The point is that the arguments go
through essentially without change, when the unknown is an N-tuple
of functions, the unperturbed operator is $(\Box + m^2)1$ and the perturbation
is a first-order matrix perturbation. Thus the propagation character of
solutions of the Dirac equation is not altered no matter how it is
perturbed by a smooth local perturbation.

This stability of the Dirac equation under smooth local perturbation
is to be contrasted with the behavior of the Petiau–Duffin–Kemmer
equations for spin zero and one.[34] They have the form when perturbed

$$[\beta^\mu \partial_\mu + m + B(x)]\psi(x) = 0, \tag{1.56}$$

where the β^μ satisfy

$$\beta^\lambda \beta^\mu \beta^\nu + \beta^\nu \beta^\mu \beta^\lambda = -g^{\lambda\mu}\beta^\nu - g^{\mu\nu}\beta^\lambda. \tag{1.57}$$

For spin zero, the βs are 5×5 and for spin one, they are 10×10, and
the same is true of B. Furthermore, B is assumed to belong to $\mathscr{S}$ as
before, and to satisfy (1.30). For spin zero, the analogue of the decomposition (1.52) is

$$B(x) = \rho(x)1 - \sigma(x)\tfrac{1}{3}(1+\beta^\rho\beta_\rho) - eiA^\mu(x)\beta_\mu$$
$$+ B^\mu(x)[-\tfrac{1}{3}(1+\beta^\rho\beta_\rho), \beta_\mu]_- + (i/2)F^{\mu\nu}(\beta_\mu\beta_\nu - \beta_\nu\beta_\mu) \tag{1.58}$$
$$+ \tfrac{1}{2}G_{\mu\nu}(x)[\beta^\mu\beta^\nu + \beta^\nu\beta^\mu - \tfrac{1}{2}g^{\mu\nu}\beta^\rho\beta_\rho].$$

The analogous expansion of B for spin one has fifteen terms.[35] They
will not be listed here.

The instability encountered here is the phenomenon, discovered by
Velo and Zwanziger,[36] of non-causality. In the presence of a suitable
external field, influence propagates faster than light; the support of the
retarded function spreads beyond the light cone and, for sufficiently
strong fields, the wave equation is no longer hyperbolic at all. The
effect of even a smooth local perturbation is to cause the particles
described by the theory to move faster than light. Expressed in another
way, what this discovery shows is that there exist equations and external
perturbations for which a weakly retarded fundamental solution, if it
exists at all, is *not* retarded. Later work has shown that if one is willing to
consider wave equations in space time of two dimensions, there are cases
in which the weakly retarded but not retarded fundamental solution
does exist.[37]

To see this phenomenon occurring for spin zero, one has only to
consider coupling to an external field $G_{\mu\nu}$, that is a symmetric traceless

tensor. The second degree polynomial that defines the characteristics turns out to depend on G in such a way as to produce motion faster than light. On the other hand, there are other couplings both for spin zero and spin one, which are stable. For example, the minimal coupling $-eiA^\mu(x)\beta_\mu$. To see this one can use a neat argument of L. Svensson.[38] The Duffin–Kemmer algebra defined by the relations (1.57) contains an element $d(a)$ such that for all complex four vectors a

$$(-ia^\mu\beta_\mu+m)d(a)=(-a^2+m^2)1. \tag{1.59}$$

In fact,

$$d(a)=(i\beta^\mu a_\mu+m)((i\beta^\mu a_\mu)/m)-((a^2-m^2)/m). \tag{1.60}$$

If one replaces a^μ by $+i[\partial^\mu-eiA^\mu(x)]$ then it turns out that (1.59) is replaced by

$$[\beta^\mu(\partial_\mu-eiA_\mu(x))+m][d(i(\partial^\mu-eiA^\mu))]$$
$$=[(\partial^\mu-eiA^\mu(x))(\partial_\mu-eiA_\mu(x))+m^2]1 \tag{1.61}$$
$$+\text{First Order Matrix Operator.}$$

This equality is precisely what is necessary for the validity of the argument used to prove the preceding theorem about fundamental solutions of the Dirac equation. Thus, unique retarded and advanced fundamental solutions also exist for the Duffin–Kemmer equation with minimal coupling and no instability occurs.

The full consequences of this instability phenomenon are not yet known. In particular, it is not yet clear whether analogous phenomena occur in theories of coupled fields. What is clear is that the Dirac equation occupies a privileged position.

2. EQUATIONS FOR LEPTONS

Since the high energy neutrino experiments of the early sixties it has been clear that the leptons occurring in Nature include at least the two electrons, $e^\pm$, the two muons, $\mu^\pm$, and the two neutrinos, ν_e, ν_μ, and their anti-neutrinos, $\bar\nu_e$, $\bar\nu_\mu$. It is a natural idea to attempt to write a wave equation which describes all these simultaneously. In the past this has been done in one of two ways. First, and trivially, one can simply write separate Dirac equations for $e^\pm$, $\mu^\pm$ and the neutrinos; that amounts to writing a reducible twelve component equation for the whole bunch. Second, following Ward and Salam[39] or Weinberg,[40] one can start by assuming they all have mass zero and account for their mass differences as an effect of interaction. Here a third way which

includes the first as a special case and can be regarded as a phenomenological parametrization of the dynamical predictions of the second
will be considered. The idea is to consider a general wave equation

$$(\beta^\mu \partial_\mu + \rho)\psi(x) = 0, \tag{2.1}$$

where β^μ and ρ transform under a representation $A \to S(A)$ of SL(2, C)
as follows

$$S(A^{-1})\beta^\mu S(A) = \Lambda(A)^\mu{}_\nu \beta^\nu \tag{2.2}$$

$$S(A^{-1})\rho S(A) = \rho. \tag{2.3}$$

The set of β^μ and ρ consistent with these transformation laws (2.2) and
(2.3) forms a vector space. For each choice of β^μ and ρ the equation
predicts a mass spectrum of particles. Fixing the masses corresponds to
choosing β^μ and ρ from a certain algebraic surface in the vector space.
The form factors associated with the equation, which are matrix
elements of the solutions for different momentum and spin, depend on
which β^μ and ρ is chosen from the algebraic surface.

One has therefore a potentially greater freedom to describe experimental properties of leptons than if, following the first method mentioned above, one simply writes separate Dirac equations for each
particle. At the moment there does not appear to be any crying need
for this additional freedom to explain experiments but it is also true
that the theoretical possibilities have not been studied. There is a
second reason for working out these ideas. That is the possibility that
the weak and electromagnetic interactions may have a succinct expression in terms of the fields following from this single particle theory.
Although an enormous effort has been put into an analogous program
in strong interaction physics (wave equations for towers of particles)
no striking successes have been achieved. Whatever the merits of such
speculations may be it seems worthwhile to understand what the possibilities are for the leptons. As will be seen the theory has new features
when it has to describe both massive and massless particles.

The first steps in the theory of (2.1) are the standard ones. Passing
to the Fourier transform one has

$$(-\not{p} + \rho)\hat{\psi}(p) = 0 \quad \text{with} \quad \not{p} = i\beta^\mu p_\mu. \tag{2.4}$$

This equation has non-trivial solutions only for p satisfying

$$\det(-\not{p} + \rho) = 0.$$

By virtue of (2.2), (2.3), and the fundamental theorem on vector invariants the left-hand side of (2.5) is a polynomial in p^2.

$$\det(-\not{p} + \rho) = \mathcal{P}(p^2) = q_0(-p^2 + m_1{}^2) \ldots (-p^2 + m_n{}^2),$$

with roots $m_1^2 \ldots m_n^2$. Since the interpretation of these roots is as the mass spectrum of the theory we require $m_1^2 \geqslant 0 \ldots m_n^2 \geqslant 0$. In order that at least one of the roots vanish we have the criterion of Daniel Kwoh.[41]

Lemma. A necessary condition that (2.4) have a zero mass solution is that

$$\det(-p + \lambda\rho) = 0,$$

hold for all real λ and all light-like p. In particular, $\det \rho = 0$ and $\det p = 0$ for all light-like p.

The criterion is a simple consequence of the fact that the Lorentz group acts transitively on the light cone.

Up to this point the discussion applies to any wave equation of the form (2.1). Now it will be specialized to the case of a twelve-component wave function with a transformation law under SL(2, $\mathbf{C}$) under a representation

$$A \to \left\{ \begin{array}{c|c|c} S(A) & 0 & 0 \\ \hline 0 & S(A) & 0 \\ \hline 0 & 0 & S(A) \end{array} \right\}, \tag{2.5}$$

where from this point on $A \to S(A)$ stands for the 4×4 transformation law of a Dirac spinor. (2.5) is the simplest transformation law compatible with known lepton states. The following discussion of the equations compatible with (2.5) is far from complete but gives a preliminary survey of the possibilities.

Since (2.5) can be written $\mathbf{1} \otimes S(A)$ where $\mathbf{1}$ stands here for the 3×3 unit matrix, and since the set of matrices commuting with $S(A)$ is spanned by the 4×4 matrices $\mathbf{1}$ and γ^5, the possible ρ satisfying (2.2) are of the form

$$\rho = \rho^{(0)} \otimes \mathbf{1} + \rho^{(5)} \otimes \gamma^5, \tag{2.6}$$

where $\rho^{(0)}$ and $\rho^{(5)}$ are arbitrary 3×3 matrices. Furthermore, since all solutions, Γ^μ, of

$$S(A)^{-1} \Gamma^\mu S(A) = \Lambda(A)^\mu{}_\nu \Gamma^\nu, \tag{2.7}$$

are linear combinations of γ^μ and $\gamma^\mu \gamma^5$, the most general β^μ is of the form

$$\beta^\mu = \beta^{(0)} \otimes \gamma^\mu + \beta^{(5)} \otimes \gamma^\mu \gamma^5, \tag{2.8}$$

$\beta^{(0)}$ and $\beta^{(5)}$ being arbitrary 3×3 matrices.

For each pair ρ, β^μ given by (2.6) and (2.8) there is a mass spectrum given by the roots of

$$\det(-\beta^{(0)}\otimes\not p - \beta^{(5)}\otimes\not p\gamma^5 + \rho^{(0)}\otimes 1 + \rho^{(5)}\otimes 1) = 0. \qquad (2.9)$$

Since we know this is a polynomial in p^2, there is no loss in generality in considering p of the special form $(p^0, 0, 0, 0)$. Then, multiplying (2.9) for convenience by $\det(1\otimes\gamma^0)$ and choosing a representation in which

$$\gamma^0 = \left(\begin{array}{c|c} 1 & 0 \\ \hline 0 & -1 \end{array}\right) \qquad \gamma^5 = \left(\begin{array}{c|c} 0 & 1 \\ \hline 1 & 0 \end{array}\right),$$

we get (2.9) in the form

$$\det[-(\beta^{(0)}\otimes 1 + \beta^{(5)}\otimes\gamma^5)p^0 + \rho^{(0)}\otimes\gamma^0 + \rho^{(5)}\otimes\gamma^0\gamma^5]$$

$$= \det\left\{\begin{array}{c|c} (-\beta^{(0)}p^0 + \rho^{(0)})\otimes 1 & (\beta^5 p^0 + \rho^{(5)})\otimes 1 \\ \hline (\beta^{(5)}p^0 - \rho^{(5)})\otimes 1 & (-\beta^{(0)}p^0 - \rho^{(0)})\otimes 1 \end{array}\right\} = 0, \qquad (2.10)$$

where in the second expression the 1s are 2×2. The last determinant is the square of the 6×6 determinant

$$\det\left\{\begin{array}{c|c} -\beta^{(0)}p^0 + \rho^{(0)} & \beta^{(5)}p^0 + \rho^{(5)} \\ \hline \beta^{(5)}p^0 - \rho^{(5)} & -\beta^{(0)}p^0 - \rho^{(0)} \end{array}\right\}. \qquad (2.11)$$

Expanded it is a polynomial of third degree in $(p^0)^2$, the constant term being

$$\det A = \det\left\{\begin{array}{cc} \rho^{(0)} & \rho^{(5)} \\ -\rho^{(5)} & -\rho^{(0)} \end{array}\right\}, \qquad (2.12)$$

the coefficient of $((p^0)^2)^3$

$$\det B = \det\left\{\begin{array}{cc} -\beta^{(0)} & \beta^{(5)} \\ \beta^{(5)} & -\beta^{(0)} \end{array}\right\}, \qquad (2.13)$$

while the coefficients of the quadratic and quartic terms are respectively the forms

$$X = \sum_P \sigma(P) \sum_{1\leqslant k_1 < k_2 \leqslant 6} A_{1P(1)} \cdots B_{k_1 P(k_1)} \cdots B_{k_2 P(k_2)} \cdots A_{6P(6)} \qquad (2.14)$$

$$Y = \sum_P \sigma(P) \sum_{1\leqslant k_1 < k_2 < k_3 < k_4 \leqslant 6} A_{1P(1)} \cdots B_{k_1 P(k_1)} \cdots B_{k_4 P(k_4)} \cdots A_{6P(6)}.$$

Thus the conditions on $\rho^{(0)}$, $\rho^{(5)}$, $\beta^{(0)}$, $\beta^{(5)}$, that guarantee the correct mass spectrum are,

$$\det A = 0,$$
$$m_e^2 + m_\mu^2 = -Y[\det B]^{-1},$$
$$m_e^2 m_\mu^2 = X[\det B]^{-1}. \qquad (2.15)$$

These equations describe the algebraic surface of admissible ρ and β^μ referred to above.

To complete the description of the admissible theories, it is necessary to analyze the structure of the conserved currents associated with the equation and to impose the requirement that one of them have the appropriate positivity properties to provide a scalar product for the space of positive energy solutions of the equation. To choose among the surviving equations, one has to analyze how a theory of electromagnetic and weak interactions could be constructed using the corresponding fields. A full description of such an enterprise would be out of place here. Its difficulty only serves to emphasize the limited character of the progress that has been made since the great days when Dirac created a theory of electrons and the electromagnetic field.

NOTES

1. P. A. M. Dirac, 'The quantum theory of the electron', *Proc. Roy. Soc.* (*London*) A117, 610–24 (1928).
2. P. A. M. Dirac, 'The quantum theory of the electron II', *Proc. Roy. Soc.* (*London*) A118, 351–61 (1928).
3. Our notation for the γs and scalar products follows Bjorken and Drell, *Relativistic Quantum Mechanics*, p. 281 (McGraw-Hill: 1964),
$$\gamma^\mu\gamma^\nu + \gamma^\nu\gamma^\mu = 2g^{\mu\nu} \quad g^{00} = 1 = -g^{ii} \quad i = 1, 2, 3.$$
We write, however, $\psi^\dagger = \bar{\psi}\eta$, where $\gamma^{\mu*} = \eta\gamma^\mu\eta^{-1}$ and $\bar{\psi}$ is the complex conjugate of ψ.
4. E. P. Wigner, 'On unitary representations of the inhomogeneous Lorentz group', *Annals of Math.* 40, 149–204 (1939).
5. Ref. 1, p. 6.
6. W. Gordon, 'Der Compton Effekt nach der Schrödingerschen Theorie', *Z. Physik* 40, 117–33 (1926).
7. O. Klein, 'Elektrodynamik und Wellenmechanik von Standpunkt des Korrespondenzprinzips', *Z. Physik* 41, 407–42 (1927).
8. The point was first established as a general theorem for particles of arbitrary spin by T. Newton and E. P. Wigner, 'Localized states for elementary systems', *Rev. Mod. Phys.* 21, 400–6 (1949), but their results were preceded by a considerable number of others in part more special (spin $\frac{1}{2}$ only) and in part more general (include effects of external fields). See, for example, R. Becker, 'Die aus der Dirac-Gleichung des Elektrons folgende Zwei-Komponenten-Gleichung', *Gött. Nach.* 39–47 (1945), and L. Foldy and S. Wouthuysen, 'On the Dirac theory of spin $\frac{1}{2}$ particle and its nonrelativistic limit', *Phys. Rev.* 78, 29–36 (1950).
9. Dirac gave an approximate expression for the bound state energies. That they are given exactly by the Sommerfeld formula was proved independently by C. Darwin, 'The wave equation of the electron', *Proc. Roy. Soc.* (*London*) A118, 654–80 (1928) and W. Gordon, 'Die Energieniveaus des Wasserstoffatoms nach der Diracschen Quantentheorie des Elektrons', *Z. Physik* 48, 11–14 (1928).

I

114 A. S. WIGHTMAN

10. The standard approach is to rewrite the second order equation as a first order system of the form $i(\partial/\partial t)x = Ax$ and then to prove that A is self-adjoint for an appropriately defined norm. For results of this type see K. Veselić, 'A spectral theory for the Klein–Gordon equation with an external electrostatic potential, *Nucl. Phys.* A147, 215–24 (1970). Some of the basic ideas involved go back to L. Schiff, H. Snyder and J. Weinberg, *Phys. Rev.* 57, 315–18 (1940).

11. O. Klein, 'Die Reflexion von Elektronen an einem Potentialsprung nach der Relativistischen Dynamik von Dirac', *Z. Physik* 53, 157–65 (1928), actually discusses a time independent field strong enough to create pairs, a case for which even modern quantum field theory does not give a satisfactory treatment. However, the paradox remains for time dependent fields, which can be treated. It suffices to switch off the external field in the distant past and future.

12. P. A. M. Dirac, 'A theory of electrons and protons', *Proc. Roy. Soc.* (*London*) A126, 360–5 (1930) as well as 'On the annihilation of electrons and protons', *Proc. Cambridge Phil. Soc.* 26, 361–75 (1931) J. Oppenheimer and H. Weyl criticized the identification of holes with protons, and Dirac accepted their criticism remarking, in an aside in 'Quantized singularities in the electromagnetic field', *Proc. Roy. Soc.* (*London*) A133, 69–72 (1931): 'It thus appears that we must abandon the identification of the holes with protons and must find some other interpretation for them. Following Oppenheimer, we can assume that in the world as we know it, all, and not nearly all, of the negative energy states are filled. A hole, if there were one, would be a new kind of particle, unknown to experimental physics, having the same mass and opposite charge to the electron. We may call such a particle an anti-electron. We should not expect to find any of them in nature, on account of their rapid rate of recombination with electrons, but if they could be produced experimentally in high vacuum, they would be quite stable and amenable to observation.'

13. W. Pauli, 'Die Allgemeinen Prinzipien der Wellenmechanik', *Handbuch der Physik* 2, Aufl. Band 24 1. Teil p. 246.

14. P. Jordan, 'Zur Quantenmechanik der Gasentartung', *Z. Physik* 44, 473–80 (1927).

15. P. Jordan and E. Wigner, 'Uber das Paulische Aquivalenzverbot', *Z. Physik* 47, 631–51 (1928).

16. V. Fock, 'Zur Theorie des Positrons', *Dokl. Akad. Nauk USSR* 1, 267–71 (1933).

17. For the required realization see, for example, R. Jost, *The general theory of quantized fields* (Am. Math. Soc.: 1965) pp. 45–50.

18. P. A. M. Dirac, 'On the infinite distribution of electrons in the theory of the positron', *Proc. Cambridge Phil. Soc.* 30, 150–63 (1934).

19. W. Heisenberg and W. Pauli 'Zur Quantendynamik der Wellenfelder', *Z. Physik* 56, 1–61 (1929); II: ibid 59, 168–90 (1929).

20. P. A. M. Dirac, 'Relativistic quantum mechanics', *Proc. Roy. Soc.* (*London*) A136, 453–64 (1932).

21. W. Pauli and V. F. Weisskopf, 'Uber die Quantisierung der skalaren relativistischen Wellengleichung', *Helv. Phys. Acta* 7, 709–31 (1934).

22. W. Pauli, *The Theory of the Positron and Related Topics* (*Report of a Seminar*), mimeographed notes The Institute for Advanced Study 1935–6.

23. W. Heisenberg, 'Bemerkungen zur Diracschen Theorie des Positrons', *Z. Physik* 90, 209–31 (1934).

24. J. J. Thomson, 'On the electric and magnetic effects produced by the motion of electrified bodies', *Phil. Mag.* **11**, 229 (1881).
25. H. A. Lorentz, *The Theory of Electrons* (Teubner: 1916).
26. P. A. M. Dirac, *Théorie du Positron*, Rapports du 7ième Conseil Solvay, pp. 203–12.
27. C. N. Yang and D. Feldman, 'The S-matrix in the Heisenberg representation', *Phys. Rev.* **79**, 972–8 (1950).
28. A. S. Wightman, 'Quasi-local and local solutions of the external field problem and their scattering theories', in preparation.
29. E. Uehling, 'Polarization effects in the positron theory', *Phys. Rev.* **48**, 55–63 (1935).
30. J. G. Valatin, 'Singularities of electron kernel functions in an external electromagnetic field', *Proc. Roy. Soc. (London)* A**222**, 93–108 (1954).
31. A. Capri, private communication.
32. L. Schwartz, *Théorie des Distributions II* (Hermann: 1951).
33. L. Hörmander, *Linear Partial Differential Operators* (Springer: 1963).
34. G. Petiau, 'Contribution à la théorie des équations d'ondes corpusculaires', *Mémoires de l'Académie Royale de Belgique* **16**, Fasc. 2 (1936); R. J. Duffin, 'On the characteristic matrices of covariant systems', *Phys. Rev.* **54**, 1114 (1938); N. Kemmer, 'The particle aspect of meson theory', *Proc. Roy. Soc. (London)* A**173** (1939).
35. A. Glass, Princeton Thesis, 1970 (unpublished).
36. G. Velo and D. Zwanziger, 'Propagation and quantization of Rarita–Schwinger waves in an external electromagnetic potential', *Phys. Rev.* **186**, 1337–41 (1969). 'Non causality and other defects of interaction Lagrangeans for particles with spin one and higher', *Phys. Rev.* **188**, 2218–22 (1969).
37. R. Minkowski and D. Seiler, 'Massive vector mesons in external fields', *Phys. Rev.* D**4**, 359–66 (1971).
38. Private communication from L. Gårding.
39. J. Ward and A. Salam, 'Electromagnetic and weak interactions', *Phys. Rev. Letters* **13**, 168–71 (1964).
40. S. Weinberg, 'A model of leptons', *Phys. Rev. Letters* **19**, 1264–6 (1967).
41. D. Kwoh, Princeton Senior Thesis, 1970 (unpublished).

7

Fermi–Dirac Statistics

Rudolf Peierls

Every undergraduate studying physics is taught that some particles obey Fermi–Dirac and others Bose–Einstein statistics. For brevity he often talks about Fermi or Bose statistics. It has become customary to call the particles fermions or bosons, and it would not occur to anyone to talk of diracons any more than one would use the term einsteinons. So our nomenclature in one sense does, in another does not, associate Dirac's name with antisymmetric statistics.

The basis for this association is one paper (Dirac, 1926) which Dirac wrote in 1926 – at least I am not aware of any other paper directly relevant to the subject. To understand the connection we therefore have to go back to 1926, a wonderful year in physics, which no doubt will find mention in other articles in this volume.

At the beginning of 1926 modern quantum mechanics had just started to emerge. Heisenberg's basic paper had appeared in September 1925 (Heisenberg, 1925). Born and Jordan (1925) elaborated in a paper published in November and in a joint paper with Heisenberg (Born, Heisenberg, and Jordan, 1926) in February 1926. Dirac (1925) had at once appreciated the new ideas and elaborated his own approach to the formalism, starting with his paper submitted in November and published in December 1925, in parallel with Born and Jordan. He continued the development with four papers published in 1926. Meanwhile Schrödinger's papers (1926a, b, c, d, e) on wave mechanics had started to appear. The first of these was published in March 1926, and the one showing the connection with Heisenberg's theory appeared in May.

On the subject of statistics, there were three different and quite independent, approaches made that year by Fermi, Heisenberg, and Dirac. Fermi's work was sent to the Accademia dei Lincei as a brief summary (1926a), and the full publication sent to *Zeitschrift für Physik* in March and published in May (1926b).

His starting point was the belief that the Third Law of Thermodynamics should apply also to a perfect gas: Fermi knew that for gas atoms in a finite 'box' the motion would be periodic, and hence have discrete energy levels, but since their spacing decreases as the size of the box increases, this would not satisfy the Third Law when applied to a macroscopic system. Fermi then noted that the Pauli exclusion principle, which had been formulated only the year before (Pauli, 1925), gave an opportunity of arriving at a different behaviour. He was evidently reluctant to work in terms of the quantized states of particles in a box, and instead introduced an oscillator potential, in which it was easier to visualize the gradual filling of states according to the exclusion principle. From this he deduced not only the distribution of particles in total energy, but also in kinetic energy, and thus obtained the formula for the equation of state of a perfect gas of particles obeying the exclusion principle, the formula now known as the Fermi gas formula.

Heisenberg's paper (1926) was sent for publication in June 1926, evidently without noticing Fermi's paper. Heisenberg considered the quantum mechanics of a system containing two identical particles, and showed that the states divided into completely symmetric and completely antisymmetric states. No transition from one type to the other could be caused by any disturbance which maintained the identity of the particles. He showed that the anti-symmetric states corresponded to those permitted by the exclusion principle, and used the results for a detailed discussion of the ortho and para states of the helium atom.

It is interesting to observe that, while the main arguments of this Heisenberg paper are naturally formulated in terms of matrix mechanics, the later sections express the situation in terms of wave functions. This illustrates again the speed with which concepts were spreading at this time, since Schrödinger's first paper had appeared only in March, and the one connecting matrix and wave mechanics in May. Heisenberg did not look at the implication of the results for the equation of state of a gas.

Dirac's paper (1926), completed at the end of August, was written independently of both Fermi and Heisenberg, though Heisenberg's paper is quoted in a note added in proof. We are here interested in the second section of the paper, which deals with systems of identical particles. Here Dirac starts from the requirement that the theory should not make statements about unobservable quantities, and therefore for two identical particles which may be in the quantum states m and n, respectively, the state m, n should not be counted separately from n, m. Only one state should appear in the matrix multiplication, and this means that the state obtained by acting with a physically meaningful quantity on a possible state of the system can be expanded into a series of such states. Since physically meaningful quantities preserve the symmetry, this condition is met by the states being either symmetric or antisymmetric.

In presenting these arguments Dirac freely uses the techniques of both matrix and wave mechanics, having in the introduction noted Schrödinger's proof of their equivalence, and in the first section presented their relation in a rather more general context.

He also remarks that the antisymmetric states are those satisfying the exclusion principle and that they are appropriate for electrons in an atom, whereas the symmetric states, leading to Bose–Einstein statistics, are applicable to photons. In one of the few remarks which later developments have shown to be somewhat naive, he goes on to conjecture that the antisymmetric (i.e. Fermi–Dirac) statistics is probably correct for molecules in a gas, 'since it is known to be correct for electrons in atoms, and one could expect molecules to resemble electrons more closely than light quanta'.

He then goes on to derive the distribution of particles over states in the case of a perfect gas of particles obeying the antisymmetry condition. This is the same result as that obtained by Fermi, but the derivation is much simpler and more direct, since it starts from the solutions of Schrödinger's equation for particles in a rectangular box.

Thus, the concept of Fermi–Dirac statistics is firmly established by the end of 1926. In reviewing its further development one must remember that the word 'statistics' is used frequently with two alternative meanings.

One of these meanings refers to the statistics of the Fermi–Dirac gas, i.e. a system of many non-interacting particles obeying the Pauli principle, either free or in some suitable potential field. The distribution of particles in such a system is then given by the function derived in the basic papers of Fermi and of Dirac.

In the other sense one asks about the 'statistics' of a certain kind of particle, meaning whether they have symmetric or antisymmetric wave functions. In this sense one does not necessarily have in mind large numbers of particles or any application of statistical mechanics. In this context the emphasis is on the symmetry considerations first presented in the papers by Heisenberg and by Dirac.

We shall first look at the further development of the latter problem. The immediate consequence of noting the antisymmetry of the electronic wave function of an atom was the remark, already contained in Heisenberg's paper, that as long as the spin-orbit coupling may be neglected, the wave function factorizes into an orbital and a spin function; for two electrons each of these may be symmetric or anti-symmetric. Overall antisymmetry therefore allows only two possibilities, an antisymmetric orbital state with a symmetric spin function ('parallel' spins) or a symmetric orbital state with an antisymmetric spin function ('opposite' spins). This classification of states, which is the distinction between ortho and para states of helium, becomes more complex for several electrons.

The complete classification of the permissible symmetry types for any number of electrons with spin, but with weak spin-orbit coupling, requires the methods of group theory, in particular the theory of the permutation group. Indeed, during the earliest period of work on the analysis of atomic spectra much explicit use was made of the representations of the permutation group, and a command of group theory was then an essential prerequisite for any student of the theory of atoms. Dirac (1929) showed how the exclusion principles could be used to replace orbital exchange by spin operators, and this method allows some of the group theoretic arguments to be replaced by elementary algebraic manipulations. A similar device due to Slater (1929) goes back to the complete antisymmetry of the state if orbit and spin are allowed for. It is a convenience to label the spin states even when spin is of no direct importance, and write a single determinant of orbital and spin states for the wave function. After this, group theory was no longer essential to deal with atomic states, though it remains of course an important tool for other purposes.

Meanwhile the question of which particles other than electrons required antisymmetric wave functions (i.e. were 'fermions') received further clarification. For a composite system of n_1 fermions and n_2 bosons, bound together in a unique internal state, it is easy to see that one obtains Bose–Einstein behaviour if n_1 is even and Fermi–Dirac behaviour if n_1 is odd. The reason is that the exchange of two such systems can be performed in stages, exchanging one of the constituents of the two systems in turn. For each pair of fermions exchanged the wave function changes sign, giving a final sign change of $(-1)^{n_1}$. This fact must have been understood very early, but it does not seem easy to find a reference to its first mention.

By this rule the statistics of molecules is reduced to the statistics of the nuclei and to the symmetry of the electronic states, and since the behaviour of the electrons is well understood in many cases the main new factor is the statistics of the nuclei.

The possibility of finding the statistics of nuclei directly appeared in the observations of band spectra of diatomic molecules made of identical atoms. Observations had shown that in some cases the lines in the rotational bands showed alternating intensities, and Mecke (1925) noticed that this behaviour was found only when the two atoms in the molecule were identical. Heisenberg (1927) showed that this pheno-menon arose from the symmetry. A rotation of the molecule by 180 degrees changes the rotational wave function by $(-1)^l$, where l is the rotational quantum number. Such a rotation is, however, also equivalent to an interchange of the nuclei, followed by a rearrangement of the electrons. Since the sign change of the wave function must be the same under the rotation and under the exchange, there is a unique connection between the rotational quantum number and the symmetry of the orbital wave function of the nuclei (the connection depending on the electronic state of the molecule).

If the nucleus is spinless, the orbital wave function has to be even or odd, according to whether the nuclei obey Bose–Einstein or Fermi–Dirac statistics. In that case only even or only odd states of l can occur; half the expected lines in the band spectrum will be missing.

If the nucleus has spin $\frac{1}{2}$, the two nuclear spins can be parallel or opposite, and the orbital wave functions odd or even. In the absence of forces acting on the nuclear spins these two classes of states will occur with frequencies in the ratio of the weights of the spin states, in this case $3 : 1$. For general spin I, the ratio is $I+1 : I$. Heisenberg showed that this explained the observed alternation in intensity and could indicate the magnitude of the nuclear spin.

Rasetti (1929, 1930) was able to observe the rotational lines in the Raman spectra of O_2 and N_2 with sufficient resolving power to deter-mine which quantum numbers belonged to the allowed or stronger lines. He deduced that both the oxygen and nitrogen nuclei obeyed Bose–Einstein statistics. This seemed to contradict the rule about composite particles, since at the time nuclei were still believed to con-sist of protons and electrons. In that case all nuclei of odd charge, including that of nitrogen, should be fermions. This result augmented the list of reasons to doubt the current ideas about nuclei. The dis-covery of the neutron suggested that the constituents were protons and neutrons. The number of fermions, which settles the statistics, is then the mass number, and the contradiction disappeared.

A related difficulty remained, however, in the beta decay, since the initial and final nuclei in this process have the same mass number, and hence the same statistics. Yet the decay was believed to involve the emission of only an electron, and this required a change in statistics. This argument was one of the reasons for Pauli suggesting the neutrino hypothesis, whose experimental confirmation thus finally settled another query about statistics.

It had now been shown that all known particles (elementary or composite) of zero or integral spin obeyed Bose–Einstein statistics, and all those with half-integral spin Fermi–Dirac statistics.

Pauli (1940) showed that this rule is not accidental, and that an attempt to write theories with the 'wrong' connection between spin and statistics would meet difficulty. It is easy to see why one cannot have bosons with half-integral spin. Relativity requires the energy density to be part of a tensor of second rank (the stress tensor) which should obey a conservation law, with the energy density being always positive. For the wave equation describing a particle of half-integral spin it is impossible to construct a conserved second-rank tensor with a non-negative (4, 4) component. The obvious way to avoid physical contradictions from transitions to states of negative energy is through the idea of Dirac's 'hole' theory, for which however the exclusion principle is essential. For bosons, filling the negative energy states with particles would not prevent the transitions.

The other part of the rule, no integral-spin fermions, is less obvious, but Pauli showed that it is then not possible to have field quantities commute with each other at points having a spacelike distance from each other, e.g. at two space points at the same time. Loss of this 'microscopic causality' relation would invalidate much of the content of present physical theory, although there does not seem to be a formal proof that a theory without this condition could not be constructed.

Since then many new particles have been discovered, starting with the pion, which has zero spin and is a boson, and no contradiction with the spin statistics rule has been found.

One might therefore regard the question as closed, but doubts still arise from time to time. One such doubt concerns the assumptions underlying Pauli's proof. Feynman (1949) showed that it was possible to develop a treatment for the case of 'wrong' relation of spin to statistics, in which the energy was non-negative, and in which fields at spacelike points commuted. This theory, however, required the use of an indefinite metric, i.e. an expression for the probability of an event which was not necessarily positive (Pauli, 1950). Many authors including

Dirac have looked at the possibility of an indefinite metric at times. Such an hypothesis raises a host of new problems, and it has not been shown that the use of such a theory is consistently possible, nor has the contrary been demonstrated.

Another doubt concerns the possibility of 'parastatistics'. This term refers to a possibility, first discussed seriously by Green (1953), in which particles of a certain class might have wave functions which could be symmetric in, say, n of these particles, but would then have to be odd under the exchange of one of these with all particles other than the n. This would be a case of parafermions of nth order. This is very similar to how one would describe the possible orbital states of electrons if one had no information about their spins. They would then indeed be like parafermions of order 2, since the exclusion principle allows two, but not more than two, electrons to be in the same orbit. The difference is, however, that parafermions are not supposed to have an additional degree of freedom, and the statistical weight of one such particle in a given state is assumed to be unity, and not n, as it would be if they were just fermions with additional unobserved, or unobservable, degrees of freedom.

Such speculations have recently received particular attention because of the theory of 'quarks', hypothetical constituents of all strongly interaction particles introduced by Gell-Mann as a particularly simple and attractive way of describing the regularities and symmetries amongst these particles.

According to this theory the group of eight particles which includes the nucleons, is composed of three quarks in their most tightly bound state. To account for the properties of such particles, the three quarks must be in a state symmetric in their spins, isospin and the remaining internal quantum number. If they were bosons, the nucleons could not be fermions. If the quarks were fermions, they would have to be in a totally antisymmetric orbital state (Greenberg, 1964). This would be a rather surprising situation, and would require binding forces between the quarks which are either strongly dependent on the internal degrees of freedom, or strongly non-local. While none of this can be ruled out, it does make the thought of parafermions of order 3 an attractive alternative.

In view of such speculations it must be left to the future to settle whether our present understanding of the possible forms of statistics is complete, or whether there remains more to be learned.

We now turn to a brief review of the development of Fermi–Dirac statistics in the other sense, as the description of the properties of a large number of weakly-interacting fermions. The first applications were to electrons, both because these were the first particles known to be fermions, and also because, being light, they show degeneracy, i.e. deviations from Boltzmann statistics, at lower densities and higher temperatures than other systems.

Pauli, whose exclusion principle was underlying the whole development, saw that the new statistics could explain the paramagnetism of normal metals. It had been puzzling why many metals, including the alkalis, which seemed to fit the simple picture of a free electron gas best, showed a strong, but temperature independent paramagnetism. The discovery of the electron spin suggested a mechanism for paramagnetism, but this would have been expected to obey Curie's law, with a susceptibility inversely proportional to the temperature.

Pauli (1927) argued that on the new statistics all low-lying states of orbital motion were occupied by two electrons of opposite spin. Turning any of these spins would violate the exclusion principle, unless the electron was also moved to a state of higher kinetic energy, not yet containing an electron of the desired spin orientation. The chance of this happening is negligible, unless the initial state of the electron is within kT (k being Boltzmann's constant and T the temperature) of the edge of the region of occupied states. Thus the number of electrons participating in the magnetization is proportional to T, and this temperature dependence cancels the T^{-1} in Curie's law. The actual reduction factor is of the order of kT/E_0, where E_0 is the 'Fermi energy', and this is consistent with the observed susceptibilities for reasonable values of the electron density.

This solution to one of the paradoxes in the electron theory of metals demonstrated that Fermi and Dirac were right in their extension of the exclusion principle to electrons not in the same atom, and perhaps spread over macroscopic dimensions. In accepting this idea one has to overcome the apparent conceptual difficulty that the principle is applied to limit the behaviour of two electrons which may be 10 cm apart, and it seems hard to believe that one of them can be affected, or can even receive information, about what the other is doing. It was soon realized, of course, that this apparent contradiction is resolved by the uncertainty principle: If the distance between the electrons is known to sufficient accuracy to exclude their being close to each other, the resulting uncertainty in their relative momentum exceeds the difference between successive momentum eigen states in the 'box' in which they are moving, and the question whether or not they are in the same state of orbital motion cannot be answered by observation.

The same point can be made by using as basis for the description of the electrons not stationary states, each spread over the whole 'box', but a series of wave packets, each specified by a certain location of its centre, and by a certain mean momentum, the extent in space and in momentum corresponding with the limits of the uncertainty relation. In that case the exclusion principle is satisfied by attributing to different electrons of the same spin orientation different wave packets, differing *either* in position *or* in momentum. For electrons in different positions the exclusion principle then, in this sense, provides no restriction to their motion.

The study of electrons in metals was extended by Sommerfeld (1928), who noticed that the new statistics could also resolve another paradox; the absence of a contribution by the conduction electrons to the specified heat of a metal. This had been particularly frustrating, since the Wiedemann–Franz number, i.e. the ratio between thermal and electric conductivity, agreed with prediction, suggesting therefore that the specific heat of the electrons was normal. On the new statistics, only electrons within kT of the Fermi surface contribute to the specific heat, and that makes the contribution small in normal circumstances; but the electrons carrying both the electric current in a field, or the heat flux in a temperature gradient, are just the ones in the border region, and *their* specific heat is normal.

These considerations moved the study of metals, and of solids in general, from a state of general bewilderment to a situation in which one could see reason, and this formed the starting point of the quantum theory of solids, which has become one of the major, and one of the more useful branches of physics.

Another important application of the Fermi–Dirac statistics for electrons came with Fowler's (1926) remark that in some stars the density of matter was high enough to cause considerable degeneracy, so that the contribution of the electrons to the energy density and pressure was far greater than the thermal values of the classical theory. This argument was extended by Chandrasekhar (1935) to cover the case of the white dwarfs, in which the Fermi energy of the electrons is highly relativistic, so that the equation of states takes yet another form.

Other applications are to the theory of nuclei, treating their constituent nucleons as a gas of fermions, and to liquid ^{3}He after the discovery of that isotope.

In most of these cases, for electrons in metals, for nucleons in nuclei, and for liquid ^{3}He, the mutual interaction between the particles seems much too strong to regard them as a gas of non-interacting objects. The good results obtained with the Fermi–Dirac gas model therefore appeared puzzling. The answer to this problem was first clearly understood by Landau and later incorporated in a formal theory (Landau, 1956). He pointed out that, because of the effect of degeneracy, the interactions in a highly degenerate Fermi–Dirac gas were inefficient in deflecting particles from their free motion, because so many of the possible final states were, in fact, forbidden. Thus even moderately strong forces only change some parameters of the particle motion, such as their effective mass, or their effective collision cross sections, but do not influence the dynamics which, apart from the changes in parameters, is still to a good approximation that of free particles. For a system of fermions meeting these conditions Landau introduced the term 'Fermi liquid', and in some sense this may be regarded as completing our understanding of Fermi–Dirac statistics. However, this a subjective choice because from this point one is led to the more general problem of systems of many fermions, and hence to refinements in the treatment of many-body systems, a complex and rapidly expanding subject.

This article started with a somewhat playful reference to the question of credit and the naming of discoveries. But the really interesting impression one derives from a brief look of the development is one of wonder at the speed and sureness (but for a few minor misinterpretations) with which the whole structure of quantum theory fell into place.

Looking back at the time around 1926 it seems almost unbelievable how rapidly a few physicists, and amongst them particularly Dirac, understood and elaborated from a few clues the main logical structure and the important consequences of the new physics. It is equally impressive how fast the understanding of new concepts spread, so that a new idea proposed in one paper would within months be treated as a familiar tool in a paper by another author. This rapid growth and dissemination of understanding took place although (or because?) the number of theoretical physicists was only a very small fraction of what it is today.

REFERENCES

Born, M., Heisenberg, W. and Jordan, P. (1926). *Z. Physik* **35**, 557.

Born, M. and Jordan, P. (1925). *Z. Physik* **34**, 858.

Chandrasekhar, S. (1935). Monthly Notices, *R. Astr. Soc.* **95**, 207.

Dirac, P. A. M. (1925). *Proc. Roy. Soc. (London)* A**109**, 642.

Dirac, P. A. M. (1926). *Proc. Roy. Soc. (London)* A**112**, 661.

Dirac, P. A. M. (1929). *Proc. Roy. Soc. (London)* A**123**, 174.

Fermi, E. (1926*a*). *Atti Accad. Lincei* **3**, 145.

Fermi, E. (1926*b*). *Z. Physik* **36**, 902.

Feynman, R. P. (1949). *Phys. Rev.* **76**, 749.

Fowler, R. H. (1926). Monthly Notices, *R. Astr. Soc.* **87**, 114.

Green, H. S. (1953). *Phys. Rev.* **90**, 270.

Greenberg, W. O. (1964). *Phys. Rev. Letters* **13**, 598.

Heisenberg, W. (1925). *Z. Physik* **33**, 879.

Heisenberg, W. (1926). *Z. Physik* **38**, 411.

Heisenberg, W. (1927). *Z. Physik* **41**, 239.

Landau, L. (1956). *J.E.T.P.* **30**, 1058.

Mecke, R. (1925). *Z. Physik* **31**, 709.

Pauli, W. (1925). *Z. Physik* **31**, 765.

Pauli, W. (1927). *Z. Physik* **41**, 81.

Pauli, W. (1940). *Phys. Rev.* **58**, 716.

Pauli, W. (1950). *Progr. Theoret. Phys.* **5**, 526.

Rasetti, F. (1929). *Nature* **123**, 757.

Rasetti, F. (1930). *Z. Physik* **61**, 598.

Schrödinger, E. (1926*a*). *Ann. Physik* **79**, 361.

Schrödinger, E. (1926*b*). *Ann. Physik* **79**, 489.

Schrödinger, E. (1926*c*). *Ann. Physik* **79**, 734.

Schrödinger, E. (1926*d*). *Ann. Physik* **80**, 437.

Schrödinger, E. (1926*e*). *Ann. Physik* **81**, 107.

Slater, J. C. (1929). *Phys. Rev.* **34**, 1293.

Sommerfeld, A. (1928). *Z. Physik* **47**, 1.

8

Indefinite Metric in State Space

W. Heisenberg

In his Bakerian lecture in 1941 Dirac has suggested that in a relativistic quantum theory, use should be made of a state space with indefinite metric.[1] This was a rather revolutionary suggestion, since in unrelativistic quantum theory the state space could always be interpreted as a Hilbert space with positive metric, and the whole probabilistic interpretation of the formalism of quantum theory rested upon this assumption. Dirac's idea has been developed in the course of years in a great number of papers by many physicists, and it cannot be the intention of the present paper to give a more or less complete historical account of this development. Only its essential steps shall be briefly described and analysed, in order to arrive at conclusions about the significance of Dirac's suggestion in the present relativistic theory of elementary particles. A few years ago a rather comprehensive survey of this whole field was given in a book by Nagy,[2] which also quotes a large part of the related literature.

There were two main reasons which had led Dirac to his suggestion. One was the fact, that in a relativistic theory the indefinite metric looked more natural from a mathematical point of view; the other was the success of Dirac's theory of holes, in which it had turned out to be rather easy, by a reinterpretation of the formalism, to get rid of the negative energy values in the theory of the electron. In a similar way Dirac hoped to get rid of the negative probabilities, which formally seemed to enter through the indefinite metric.

The first point can be seen from the Klein–Gordon equation for particles of spin zero, where the charge density – contrary to Dirac's theory of spin $\frac{1}{2}$ particles – is indefinite; and from the commutator in quantum electrodynamics

$$[A_\mu(x)A_\nu(x')] = ig_{\mu\nu}D(x-x'), \tag{1}$$

where the opposite sign of g_{00} against g_{ii} indicates the indefinite metric in the corresponding state space. A more fundamental reason for this situation may be seen in the fact, that, disregarding the translations, the Lorentz-group – contrary to the Galilei-group – is a non-compact group. Any finite representation of a non-compact group requires a space with indefinite metric; though infinite representations with definite metric may exist.

The second point is more problematic. Briefly after Dirac's first paper Pauli had taken up Dirac's ideas with the intention of developing a general consistent interpretation of the new formalism.[3] This however did not seem possible, the negative probabilities could not generally be avoided; what remained were rather arbitrary rules which could not be fitted into a simple convincing scheme.

The next decisive progress was achieved in the well known reformulation of quantum electrodynamics by Gupta[4] and Bleuler[5]. These authors, starting from Maxwell's equations, treated the four components A_μ on equal footing and consequently introduced – besides the transversal photons – longitudinal and scalar photons; the latter according to (1) had a negative norm in the state space. A subsidiary condition, however, corresponding to the Lorentz condition $(\delta A_\mu/\delta x_\mu)=0$ in classical theory, prevented the longitudinal and scalar photons from appearing as free particles. They played their rôle only in the region of interaction, they were in fact responsible for the Coulomb-forces between charged particles, but they never appeared in the asymptotic region, and therefore not in the S-matrix. Thereby a new idea was introduced into quantum field theory: the total state space should be divided into two parts. The one contains all asymptotic states, which are physically allowed, and in this subspace the metric is positive. The other part, in which the metric can be indefinite, appears physically only in the interaction. The subsidiary condition, which projects the physical states out of the complete state space, guarantees the consistency of the partition independent of time, and the probabilistic interpretation of the S-matrix does not encounter any difficulties. Hence it is not surprising that the Gupta–Bleuler formalism can be transformed back into the Dirac–Schwinger formalism; it is mathematically equivalent to a theory with positive metric, where a long range interaction, the Coulomb force, has been introduced from the beginning. The indefinite metric appears here as a kind of luxury allowing a formulation of quantum electrodynamics which is manifestly covariant and contains only local interactions.

This idea of the two parts of state space and of the indefinite metric was taken up in the author's attempt for a unified field theory,[6] in order to formulate a consistent scheme which from the beginning does not lead to any infinities, divergent integrals etc. It was assumed that the two-point functions, on account of the interactions, do not contain dangerous singularities at the light cone. In the simplest approximation the two-point function for fermions was represented in momentum space by a function, that contained a pole at an average baryon mass and a dipole at mass zero. This latter assumption implied the introduction of an indefinite metric in state space, and the mass zero was chosen to include the leptons. The baryon function should be regularised by means of the leptons, the lepton function by the baryons. It was of course not clear at that time, in which way the dipole at mass zero could be connected with the leptons. The subsidiary condition, which should project the physical states out of the total state space, was formulated as the postulate, that only combinations of eigenstates of the total energy can appear as physical states.

A sufficient clarification of the dipole case and the corresponding subsidiary condition was obtained shortly afterwards by means of the Lee model.[7] Pauli and Källén had demonstrated, that for a sufficiently high (or infinite) cut-off energy the Lee model requires a state space with indefinite metric.[8] The parameters can be chosen such that in the eigenvalue problem for the 'V-particle' the two roots coincide and a dipole results. In this situation two states appear with norm zero, but not orthogonal to each other. The one (the 'good' ghost) is an eigenstate of the Hamiltonian, the other (the 'bad' ghost) is not. Solutions can be found in which only the good ghosts, not the bad ones, occur as incoming or outgoing particles, and the resulting S-matrix is unitary. Since even the good ghosts have the norm zero, they do not contribute to any expectation value, hence the V-particles are never created as physical states. The subsidiary condition can be formulated as indicated above; the physical states are those that can be constructed as combinations of eigensolutions of the Hamiltonian.

A similar situation arises, as could be shown by Pauli,[9] when the two roots in the eigenvalue equation of the V-particles are complex. At least in the low sectors of the Lee model (implying only one V-particle) the unitarity of the S-matrix is not endangered by such complex poles. On the other hand the case of the two real roots (one of which then belongs to a negative norm) seemed to lead to a non-unitary S-matrix.

These results gave rise to a number of papers concerning the general conditions under which a theory with a state space of indefinite metric can find a probabilistic interpretation. For the details we refer to Nagy's book.[2] A general argument of Sudarshan seemed to show that a fairly large class of theories of this kind should allow such an interpretation:[10] If incoming and outgoing waves can be defined, and an S-matrix that transforms from the one to the other, then it should be possible by a linear transformation to bring this S-matrix into its diagonal form. Some of the diagonal elements will then be of the type $e^{i\alpha}$, i.e. have the absolute value unity, some others won't. The states belonging to the first group can be defined as the physical states, the others as the rest. The unitarity of the S-matrix is then guaranteed by this definition. But of course the unitarity of the S-matrix is not the only condition which must be fulfilled by a reasonable quantum field theory.

So far the indefinite metric and the concept of the two state spaces had been used for two quite different purposes. In the Gupta–Bleuler method it had led to a formulation of quantum-electrodynamics which is manifestly covariant and avoids the non-local long-range interactions; in the Lee model and – possibly – in the unified field theory it had permitted a consistent theory of an interaction without (i.e. with an infinite) cut-off without infinities or divergencies.

A connexion between the two aspects is suggested by the conjecture, that the equivalence between the Gupta–Bleuler method and the Dirac-Schwinger formalism is not limited to quantum electrodynamics, that in modified form it plays an essential rôle in much more general cases and especially in the unified field theory of elementary particles. This conjecture has found the support of two recent papers, the one by Karowski,[11] the other one by Dürr and Rudolph.[12]

Karowski has been able to show that the Lee model, e.g. its dipole version, is mathematically equivalent to another model, in which only the N- and θ-particles, not the V-particles, are introduced from the beginning; they are connected by a non-local interaction of a special type, and the metric in state space is positive. Both models lead to exactly the same S-matrix for collisions between one N- and one or two θ-particles (the results can probably be extended to any number of θ-particles); the same analytical functions occur in both models. The V-particle appears in the second model only as bound state between N- and θ-particle, not as primary field, and this state is characterised by a pole, not a dipole. The norm connected with this pole is positive, as it should be in a state space with positive metric. Analytically this change is brought about by a factor in the eigenvalue equation for the bound state (as compared with that of the V-particle in the first model), which vanishes at the critical energy. Hence in this second model a genuine bound state of N and θ exists, with positive norm. But the analytical structure provides a selection rule which prevents the formation or annihilation of this bound state in any collision process between N- and θ-particles.

It is very tempting to use this mathematical structure as a model for the interpretation of the two-point function mentioned above, which as a first approximation had been tried in the unified field theory. The leptons had been represented by a dipole at mass zero. The system could possibly be mathematically equivalent to another one with a positive metric, not manifestly covariant and with a non local interaction. In this system the dipole at mass zero would automatically be replaced by a pole, the leptons would become genuine particles with positive norm, but the analytical structure would provide a selection rule, which in this approximation prevents the creation or annihilation of leptons, in agreement with the observations.

The close analogy between the Bleuler–Gupta method and the dipole case of the Lee model was further underlined in the paper by Dürr and Rudolph.[12] If in the Bleuler–Gupta formalism the wavefunctions of the longitudinal and the scalar photons are replaced by their sum and their difference, then the sum can be compared with the field of the good ghosts, the difference with that of the bad ghosts. The norm of both kinds of ghosts is zero, but their wavefunctions are not orthogonal to each other. The subsidiary condition, as in the Lee model, prevents the bad ghosts from appearing as free particles, while the good ghosts are permitted, but do not contribute to any expectation-value. Actually in electrodynamics the addition of good ghosts means just a change of gauge, which does not affect the expectation values of physical observables. There is one characteristic difference however: While in electrodynamics the contribution of the good ghosts is quite arbitrary, i.e. any change of gauge is permitted, this contribution is fixed in the Lee model by the condition, that the bad ghosts must not appear as incoming or outgoing waves. But aside from that the analogy is complete.

The Gupta–Bleuler method has been extended by Dürr and Rudolph to fields of higher spin,[13] especially to gravitation. Again a manifestly Lorentz invariant and local formulation of the theory requires a state space with indefinite metric. The partition of the complete state space into a subspace of physical states and the rest is again reflected in the appearance of 'good' and 'bad' ghosts. In the S-matrix the bad ghosts are eliminated by the subsidiary condition.

So far the analysis of the Gupta–Bleuler method and the Lee model has led to the conclusion that in principle a consistent relativistic quantum theory of interaction can be constructed by making use of a state space with indefinite metric, as Dirac had suggested. The state space must be divided into two parts, the subspace of the asymptotic physical states and the rest. The physical states are defined by a subsidiary condition. The 'non physical' states play their rôle only in the region of interaction; they represent – or produce indirectly – a non local interaction which is not 'quantised', in the sense that it is not connected with free particles which would be the quanta belonging to this field. In this sense neither the Coulomb field nor the Newtonian gravitation are quantised, they may be called 'classical fields' with sufficient caution.

These results have strengthened the confidence in the indefinite metric in state space, and recently in many papers this method has been applied to problems which apparently could not be solved by conventional methods. We mention the electromagnetic mass splitting in isospin multiplets and similar problems. Arons, Han and Sudarshan[14] and Lee and Wick[15] have attempted to develop quantum electrodynamics as a theory without infinities by using from the beginning the indefinite metric in state space. Both papers follow the general idea of Pauli and Villars, that heavy masses may be used to regularise the two-point functions of the electrons. Arons, Han and Sudarshan consider real, Lee and Wick, complex poles. The two papers differ by the formulation of the subsidiary condition. Lee and Wick assumed, that in the case of complex poles a simple change in Feynman's well known prescription for the path of integration should be sufficient to define the physical states. This assumption has been criticised by Sudarshan et al.[16] It seems that in higher approximations one cannot get around a careful analysis of the asymptotic behaviour of the various 'ghosts'.

This question is already closely connected with the problem of causality. A theory with indefinite metric in state space will certainly not obey those strict criteria of causality, which had been derived from the conventional axioms; but such a theory may still be compatible with the existing observations.

Recently the consequences of the indefinite metric for the problem of causality and for the analytic behaviour of the amplitudes have been discussed in several papers by Lee and Wick,[15] Sudarshan et al.[16] and Dürr and Seiler.[17]

The problem of causality is not so well defined in quantum theory as it had been in classical theory. In the latter, one could construct sharp wave fronts, and causality required that these wave fronts never travelled faster than with the velocity of light. In quantum theory it is not possible to construct such a wave front, since it would require the superposition of plane waves of all frequencies, negative as well as positive. But even for the photon field a coherent mixture of many photon states covers only the positive energy range, i.e. contains only positive frequencies. Hence the forward edge of any wavefront will not be sharp; it will, from its maximum value in the forward direction decrease exponentially or possibly by a power law or otherwise, and the form of this forward tail will depend on the analytical structure of the amplitudes as a function of energy and momentum. From the existing experimental evidence on the time order of events it will be difficult to tell which form of the tail is still permissible. But it may be possible to observe, directly or indirectly, the analytical structure of the amplitudes.

Taking the dipole case as an example it is easily seen that the amplitudes (or related functions) cannot be analytic at a possible threshold for the creation of ghost-states, because the boundary conditions must be different below and above the threshold. The amplitudes will be piece-wise analytic functions, which are continuous but not analytic at the threshold. Examples for such functions have been given in the non-linear spinor theory.[18] The consequences for the form of the 'acausal' precursors have been discussed by Dürr and Seiler.[17]

This then is the decisive new feature introduced by Dirac's indefinite metric in state space. The amplitudes will possibly not be global analytic functions – as one might have expected for a theory with purely local interaction and positive metric – but only piece-wise analytic functions, which are connected continuously at the threshold points for 'ghost'-production. Acausal precursors need not decrease exponentially; even a decreasing power law may be in agreement with the observations.

In the thirty years that have elapsed since Dirac's first paper on the indefinite metric in state space, the subject should have been clarified sufficiently to draw a conclusion about its applicability as a basis for the theory of elementary particles. The alternative, a theory with positive metric and purely local interaction, has so far not been developed in a convincing way; the author believes that the efforts in this direction are largely based on wishful thinking. Even if this could be demonstrated, it does not prove that the theory with indefinite metric can be built up to a consistent mathematical scheme. Since the mathematical problems involved are extremely difficult, it would probably be best to look for

136   W. HEISENBERG

experimental verifications of the indefinite metric. The analytical behaviour of the amplitudes in the S-matrix or of the kernels to eigenvalue equations, is at least in principle open to experimental investigation. Therefore the recent efforts of Lee and Wick to obtain experimental evidence on the problem point in the right direction. For the time being we have to be satisfied with the statement that all existing observations seem to be compatible with Dirac's hypothesis of the indefinite metric in state space.

REFERENCES

1. P. A. M. Dirac, *Proc. Roy. Soc. (London)* A180, 1 (1942).
2. K. L. Nagy, *State Vector Spaces with Indefinite Metric in Quantum Field Theory* (Akademiai Kiadó, Budapest: 1966).
3. W. Pauli, *Rev. Mod. Phys.* 15 (3), 175 (1943).
4. S. N. Gupta, *Proc. Phys. Soc. (London)* A LXIII, 681 (1950).
5. K. Bleuler, *Helv. Phys. Acta* XXIII, 567 (1950).
6. W. Heisenberg, *Z. Naturforsch* 9a, 292 (1954); *Z. Physik* 144, 1 (1956).
7. W. Heisenberg, *Nucl. Phys.* 4, 532 (1957).
8. W. Pauli and G. Källén, *Math. Phys. Medd.* 30, 7 (1955).
9. W. Pauli, *Proceedings of the Annual International Conference on High-Energy Physics*, CERN, p. 127 (Geneva: 1958).
10. E. C. G. Sudarshan, *Phys. Rev.* 123, 2183 (1961).
11. M. Karowski, *Z. Naturforsch* 24a, 510 (1969).
12. H. P. Dürr and E. Rudolph, *Nuovo Cimento X*, 62A, 411 (1969).
13. H. P. Dürr and E. Rudolph, *Nuovo Cimento X*, 65A, 423 (1969).
14. M. E. Arons, M. Y. Han and E. C. G. Sudarshan, *Phys. Rev.* 137, 4B, B1085 (1965).
15. T. D. Lee and G. C. Wick, *Phys. Rev. D*, 2, 6, 1033 (1970).
16. A. M. Gleeson, R. J. Moore, H. Rechenberg and E. C. G. Sudarshan, *Analyticity, Covariance and Unitarity in Indefinite Metric Quantum Field Theories*, CPT-77, AEC-26.
17. H. P. Dürr and E. Seiler, *Nuovo Cimento X*, 66A, 734 (1970).
18. W. Heisenberg, *Introduction to the Unified Field Theory of Elementary Particles* (J. Wiley and Sons, New York: 1966).

9

On Bras and Kets

A COMMENTARY ON DIRAC'S MATHEMATICAL
FORMALISM OF QUANTUM MECHANICS

J. M. Jauch

The general lesson of the role that mathematics has played through the ages in natural philosophy is the recognition that no relationship can be defined without a logical frame and that any apparent disharmony in the description of experiences can be eliminated only by an appropriate widening of the conceptual framework.

Niels Bohr

1. INTRODUCTION

In the preface to the first edition of his celebrated book on *The Principles of Quantum Mechanics* Dirac writes: 'A book on the new physics, if not purely descriptive of experimental work, must be essentially mathematical. All the same the mathematics is only a tool and one should learn to hold the physical ideas in one's mind without reference to the mathematical form . . .' A little further on, he says:

With regard to the mathematical form in which the theory can be presented, an author must decide at the outset between two methods. There is the symbolic method, which deals directly in an abstract way with the quantities of fundamental importance (the invariants, etc. of the transformations) and there is the method of coordinates or representations, which deals with sets of numbers corresponding to these quantities.

As regards the first or symbolic method, he continues:

The symbolic method seems to go more deeply into the nature of things. It enables one to treat the physical laws in a neat and concise way, and will probably be increasingly used in the future as it becomes better understood and its own special mathematics gets developed. For this reason I have chosen the symbolic method, introducing the representatives later on as an aide to calculation.

Although this point of view has been maintained through all editions of his book, it seems nevertheless that Dirac departed from it in his later years. In an article published in 1962 on the occasion of Wigner's sixtieth birthday he expresses the opinion that *unitary* equivalence is not necessarily the same as *physical* equivalence.[1] He says in fact:

To bring in interaction, one must depart from the point of view of looking at two representations as equivalent if they are connected by a unitary transformation, a point of view of looking upon all unitary transformations as trivial. To a physicist some unitary transformations are trivial, whereas others (for example the S-matrix) are far from trivial, so a physicist cannot look upon two representations connected by a unitary transformation as necessarily equivalent.

This second point of view has never been implemented and in the work of most theoretical physicists it is the first which has prevailed. In fact the elegant and powerful formalism of Dirac's mathematical foundation of quantum mechanics is intimately fused with the first point of view and this formalism has become practically second nature for all who work with quantum mechanical methods.

Dirac himself is fully aware that this formalism is by no means mathematically rigorous and he warns us at several places that it must be used with a certain caution in order to avoid contradictions. These warnings are particularly pertinent with respect to such typical innovations as the bra and ket vectors, the δ-function, the expansion 'theorem', the eigenbras and eigenkets, etc., etc.

There is no doubt that Dirac himself attached considerable importance to his formalism of quantum mechanics since in the book almost 80 pages out of the 300 and odd pages are devoted to its exposition. It is also evident that it has been reworded several times in subsequent editions, and that he has been aware that the mathematical rigour could and should be improved. So he says for instance in the third edition after the introduction of the δ-function:

$\delta(x)$ is not a function of x according to the usual mathematical definition of a function, which requires a function to have a definite value for each point on its domain, but is something more general which we may call an 'improper function' to show up this difference from a function defined by the usual definition. Thus $\delta(x)$ is not a quantity which can be generally used in mathematical analysis like an ordinary function, but its use must be confined to certain simple types of expressions for which it is obvious that no inconsistency can arise.

A little later on, he continues:

The use of improper functions does not involve any lack of rigour in the theory, but is merely a convenient notation, enabling us to express in a concise form certain relations which we could, if necessary, rewrite in a form not involving improper functions, *but only in a cumbersome way which would tend to obscure the argument* [italics mine].

Thus, it seems that Dirac was well aware of the work of von Neumann on the Hilbert space formulation of quantum mechanics although he never mentions him. He must also have known the work of Schwartz and others on the theory of distributions which made the δ-function and other distributions respectable. But the last quoted passage indicates that for him the mathematical rigour thus accomplished tends to obscure the argument which for him is based on analogy with the formalism in a finite-dimensional vector space.

In this context, it is very revealing that Dirac is reluctant to admit the infinite-dimensionality of the vector space with which he works. For instance in the second edition, when he introduces the vector space in paragraph 5, he introduces only 'a vector space with a *sufficiently large number of dimensions*' without specifying what this means. Only later in paragraph 10 we learn that 'we must now go over to the case of an infinite number of dimensions', in connection with the so-called 'expansion theorem'. From the context, one might be tempted to guess that infinity means uncountably infinite and a later remark seems to confirm this view. This remark occurs in the third edition, in paragraph 10, when he says: 'The bra and ket vectors that we now use form a more general space than a Hilbert space.' Since Dirac maintains the scalar product for this more general space this could only mean that he speaks here of a non-separable Hilbert space. However, Dirac avoids explicitness on this point and leaves us free to interpret his various remarks in the light of the contexts.

Since von Neumann's original book, there have been various attempts to reinterpret some of the heuristic manipulations in Dirac's formalism. There are perhaps two major lines along which one may meet the challenge presented by Dirac's formal kind of mathematics.

One of these is to extend Hilbert space in some way or other so that the various manipulations become meaningful. Such is the approach which has led to the use of nuclear spaces and distribution theory.

A second possibility is to elaborate von Neumann's Hilbert space formulation of quantum mechanics so as to make it more adapted to the actual calculational habits of physicists.

We shall in this paper develop the second approach. It is more conservative and in a sense more elementary than the first. Furthermore, by a judicious use of some less known results, it is possible to develop a calculation technique which has the twin advantage that it duplicates in almost all essential points the formal manipulations of Dirac and is, in addition, mathematically perfectly rigorous. Thus it escapes the principal objection of Dirac against mathematical rigour, viz. that it is *cumbersome and tends to obscure the argument.*

The strategy to be followed is essentially this: the theory to be developed in Hilbert space is to imitate as closely as possible the theory in finite dimensions. This is also Dirac's method, only he extends the analogy to concepts where it does not hold, with the resultant strain on the mathematical meaning. It is therefore essential to use the analogy for more convenient concepts.

Let us illustrate this with a first example, the spectrum Λ of a self-adjoint linear operator A in a Hilbert space $\mathcal{H}$. In finite dimensions the spectrum is defined as the set of real numbers λ which satisfy an equation

$$Af = \lambda f \quad f \in \mathcal{H}, f \neq \theta \text{ (zero vector).} \tag{1}$$

If $\phi_1, \phi_2, \ldots \phi_n$ is an orthonormal base in $\mathcal{H}$ the set of numbers λ are the solutions of the secular equation

$$|a_{rs} - \lambda \delta_{rs}| = 0, \tag{2}$$

where

$$a_{rs} = (\phi_r, A\phi_s) \tag{3}$$

are the matrix elements of the operator A with respect to the co-ordinate system $\{\phi_r\}$ and δ_{rs} is the Kronecker symbol

$$\delta_{rs} = \begin{cases} 1 & \text{for } r = s \\ 0 & \text{for } r \neq s. \end{cases}$$

The equation (2) is a polynomial of degree n in the variable λ and the solutions (counted with their possible multiplicities) are a set of n real numbers $(\lambda_1, \lambda_2, \ldots \lambda_n)$ called the *spectrum* Λ of A.

If one tries to transfer equation (1) into the infinite-dimensional Hilbert space, one is faced with the problem of the continuous spectrum. This may be a segment of the real axis or an even more complex subset of the reals. The continuous spectrum is physically quite important especially for the quantum theory of scattering. So it is important to have a definition of the spectrum which is also valid for the continuous spectrum. The generalization of equation (1) to infinite-dimensional Hilbert space is not possible for λ in the continuous spectrum. For such λs there are no elements in Hilbert space which satisfy this equation.

The problem then is the following: to cast condition (1) into another form which in finite dimensional spaces is completely equivalent to that condition. Furthermore this new form must be such that it can be taken over into the infinite-dimensional space without any substantial change.

The solution of this problem is found in the remark that equation (1) says that the equation

$$Xf = \theta \quad f \in \mathcal{H}, f \neq \theta, \tag{4}$$

admits a solution, where $X = A - \lambda \cdot I$ and I is the unit operator.

Now in a finite dimensional space we say an operator X is *invertible* if and only if the equation

$$Xf = g \tag{5}$$

has exactly one solution f for any value $g \in \mathcal{H}$. It is easily verified that this is the case if and only if the equation

$$Xf = \theta \quad \text{(zero vector)}$$

has only the trivial solution $f = \theta$.

This property gives us a new way of defining the spectrum of the operator A:

Let us agree to call the *resolvent set* $\rho(A)$ the set of all (complex) numbers $z \in \mathbb{C}$ for which the operator $A - z \cdot I$ is invertible. For such values we define the resolvent operator

$$R_z = (A - z \cdot I)^{-1} \quad z \in \rho(A). \tag{6}$$

It is by definition a linear operator defined on all vectors $g \in \mathcal{H}$. We have therefore the following definition:

Definition: The spectrum Λ is the set of all values $z \in \mathbb{C}$ for which $A - z \cdot I$ is *not* invertible, that is the complementary set of $\rho(A)$ in $\mathbb{C}$.

For finite dimensional spaces this definition is completely equivalent with the definition by equation (1), as can be easily verified. The advantage of using it lies in the fact that it can be generalized without change to the infinite-dimensional Hilbert space.

The only point where a little care is needed is in the interpretation of the notion of *invertible*. We need an inversion operator which is defined on *all* elements of $\mathcal{H}$. Such an operator is necessarily a bounded operator. Thus we must define invertibility in Hilbert space as *bounded invertibility*.

This simple example illustrates the principle to be followed. We shall now examine successively the concepts and formalisms as they appear in the first part of Dirac's book and translate them insofar as possible into a rigorous and consistent mathematical language.

2. BRA AND KET VECTORS

In Hilbert space $\mathcal{H}$ there is defined a scalar product, which mathematically is a function from $\mathcal{H} \times \mathcal{H}$ to the complex numbers $\mathbb{C}$ with the properties

$$(f, g_1 + \alpha g_2) = (f, g_1) + \alpha (f, g_2) \quad \forall \alpha \in \mathbb{C}$$
$$(f, g) = (g, f)^* \tag{7}$$
$$\|f\|^2 \equiv (f, f) > 0 \quad \text{for} \quad f \neq \theta.$$

Since the scalar product is written in the form of a bracket and it involves two vectors, it is suggestive to call one of them a ket and the other a bra vector. Dirac never defines what a ket vector is mathematically, he describes its physical interpretation as 'vectors which are connected with the states of a system in quantum mechanics'.[2] From the context one concludes that they are vectors in a Hilbert space (of finite or infinite dimensions) over the complex numbers.

The bras are defined in paragraph 6 on page 18 as the linear functionals over the kets.

These linear functionals are denoted by ϕ, but nothing is said about whether they are bounded or not. However, the following remark occurs concerning these functionals: '. . . the number ϕ corresponding to any $|A>$ may be looked upon as the scalar product of that $|A>$ with some new vector, there being one of these new vectors for each linear function of the ket vectors $|A>$'.

In view of Riesz' theorem on bounded linear functionals in Hilbert space, it is reasonable to interpret this remark as meaning that Dirac wants the ket vectors to represent *bounded* linear functionals.

Thus we give the following two definitions:

Definition 1: The ket vectors are the vectors in a (finite or infinite-dimensional) Hilbert space $\mathscr{H}$.

Definition 2: The bra vectors are the bounded linear functionals over $\mathscr{H}$.

To be more precise: A bra vector is a function $\phi : \mathscr{H} \to \mathbb{C}$ which satisfies the properties

$$\phi(f + \alpha g) = \phi(f) + \alpha \phi(g) \quad \forall f, g \in \mathscr{H}, \ \forall \alpha \in \mathbb{C}$$
$$|\phi(f)| \leqslant K \|f\|, \tag{8}$$

where K is some real constant.

Theorem (Riesz): Every bounded linear functional $\phi : \mathscr{H} \to \mathbb{C}$ is of the form

$$\phi(f) = (g, f)$$

with some $g \in \mathscr{H}$. The space of bounded linear functionals, also called the dual space $\mathscr{H}^*$, is thus identical with the original space. In other words, $\mathscr{H}$ is self-dual. However, the correspondence $\phi \to g$ is antilinear, since to the functional $\lambda \phi$ ($\lambda \in \mathbb{C}$) corresponds the vector $\lambda^* g$.

In view of this theorem Dirac's hypothesis on page 20 that there should exist a one-one correspondence between the bras and kets becomes superfluous, since this is precisely the content of Riesz' theorem.

We must remark, however, that if the linear functionals are unbounded then there is no such correspondence and the hypothesis is in fact false. Thus this hypothesis is consistent only with our interpretation that Dirac's bra vectors are the *bounded* linear functionals on $\mathcal{H}$.

3. LINEAR OPERATORS

Linear operators are introduced in chapter II, page 23, and they are defined as linear functions from $\mathcal{H}$ into $\mathcal{H}$. Let B be such a function:

$$B : \mathcal{H} \to \mathcal{H} \quad Bf = g \quad \forall f \in \mathcal{H}$$

satisfying the rules

$$B(f_1 + \alpha f_2) = Bf_1 + \alpha Bf_2 \quad \forall f_1, f_2 \in \mathcal{H}, \quad \forall \alpha \in \mathbb{C}.$$

On every vector $h \in \mathcal{H}$ one can then define another linear operator B^* by the rule $h^* = B^* h$

$$(h, Bf) = (h^*, f), \quad \forall f \in \mathcal{H}. \tag{9}$$

This h^* always exists because of Riesz' theorem. Furthermore the correspondence $h \to h^* = B^* h$ is linear and $(B^*)^* = B$ as one verifies without effort. The operator B is called self-adjoint if $B = B^*$.

If B_1 and B_2 are two operators then one defines sums and products by the rules

$$(B_1 + B_2)(f) = B_1 f + B_2 f$$
$$(B_1 B_2)f = B_1(B_2 f) \tag{10}$$

and one verifies readily

$$(B_1 B_2)^* = B_2{}^* B_1{}^*. \tag{11}$$

So far we have interpreted and rewritten in conventional notation the content of sections 7 and 8 with the exception of the product $|A> <B|$, which we postpone for a while, since it is a construction of a different kind.

Concerning this part, we must first make an important remark. The definitions as given here can only be used for *bounded* operators. Since in the later applications of the theory the operators are often unbounded these definitions are inadequate. The difficulty arises from the fact that unbounded operators are never defined on the entire Hilbert space but at most on a dense linear manifold. This requires some explanations:

let us agree to use the following notations and definitions. For any $f \in \mathscr{H}$ we define the unit vector $\phi = (1/||f||)f$, where $||f|| = (f,f)^{\frac{1}{2}}$. The set of all unit vectors shall be denoted by S (for Sphere). Let us consider the set of positive real numbers $||B\phi||$ as ϕ runs through S. This set may or may not have a finite supremum. In any case we *define*

$$||B|| = \sup_{\phi \in S} ||B\phi|| \tag{12}$$

and we say B is *bounded* if $||B|| < \infty$. Otherwise B is said to be *unbounded*. A bounded linear operator is also called continuous.

It is immediately obvious that every linear operator in a finite dimensional space is bounded. In Hilbert space this is not so. However the following theorem gives a sufficient condition for boundedness.

Theorem: A closed linear operator defined on all of $\mathscr{H}$ is necessarily bounded.

The corollary that an unbounded operator may not be defined on all of $\mathscr{H}$ was mentioned above.

It is usually possible to define an unbounded operator on a dense linear manifold $D \subset \mathscr{H}$. This is a linear subset, so that

$$f, g \in D \Rightarrow f + \alpha g \in D \quad \forall \alpha \in \mathbf{C}$$

and if $h \in \mathscr{H}$ then for any $\epsilon > 0$ there exists a vector $f \in D$ such that $||h - f|| < \epsilon$.

An example of such a subset is obtained for instance as follows: Let $\{\phi_r\}$ $(r = 1, 2, \ldots)$ be a complete orthonormal system of vectors in $\mathscr{H}$ (in other words a coordinate system) then the set of all vectors

$$f = \sum_{r=1}^{n} x_r \phi_r \quad x_r \in \mathbf{C}, \quad n < \infty \tag{13}$$

are a dense linear manifold in $\mathscr{H}$.

The trouble is, a dense linear manifold is a very small subset of vectors of the entire space notwithstanding the fact that they are a dense set of vectors. One can visualize this for instance on the subset of the rational numbers in the continuum. They are dense but they are countable while the continuum is not.

It should therefore not be surprising that the dense linear manifolds can be placed in many different ways into the complete Hilbert space even in such a manner that two different manifolds have only the zero vector in common.

Thus, if B_1 and B_2 are two unbounded operators and D_1, D_2 are their respective domains of definition, then the operator $B_1 + B_2$ is only defined on their intersection $D_1 \cap D_2$ and this, as we have just seen may be a very small or even trivial subset of $\mathscr{H}$.

A similar difficulty arises with the product of two such operators. The product of two operators B_1 and B_2 is only defined on the set of vectors f which satisfy the two conditions

$$f \in D_2 \quad \text{and} \quad B_2 f \in D_1.$$

Again this may be a proper or even a trivial subset of $\mathscr{H}$.

A final word on the operator $|A> <B|$. Let ϕ and ψ be two normalized vectors of $\mathscr{H}$ then we can define a linear operator X by setting

$$Xf = \phi(\psi, f). \tag{14}$$

This operator is evidently linear and bounded since

$$\|Xf\| = |(\psi, f)| \leqslant \|f\|.$$

Furthermore its adjoint is easily seen to be given by

$$X^* g = \psi(\phi, g). \tag{15}$$

If we associate ϕ with the ket $|A>$ and ψ with the bra $<B|$ then the operator X is exactly Dirac's operator

$$X = |A> <B|. \tag{16}$$

The operators of this type are called operators of rank 1. They can be used as building blocks of many other operators. In particular the so called *compact* operators are always (possibly infinite) linear combinations of rank 1 operators. However, self-adjoint operators with continuous spectrum are never of this form.

4. EIGENVALUES AND EIGENVECTORS

Dirac restricts the theory of eigenvalues and eigenvectors to self-adjoint operators (called real operators by him) since, as he says, the theory is not of much use for more general operators.

This definition of eigenvectors and eigenvalues of a self-adjoint operator is based on the equation (1)

$$Af = \lambda f \quad f \in \mathscr{H}, \quad f \neq \theta$$

mentioned in the introduction, which is not suitable for a general definition of the spectrum. We should mention, however, that equation (1) may very well have solutions. The corresponding eigenvalues are in general a subset Λ_d of the entire spectrum. Thus $\Lambda_d \subset \Lambda$. This discrete part of the spectrum may be the entire spectrum or it may be a non trivial proper subset or it may be empty.

The set of all eigenvectors f spans a linear subspace $\mathcal{H}_d \subset \mathcal{H}$. If P_d is the projection on to this subspace, then P_d reduces A, this means it commutes with A. If A is bounded, we may write this in the form

$$P_d A = A P_d. \tag{17}$$

The orthogonal complement of $\mathcal{H}_d$ is another subspace which we shall denote by $\mathcal{H}_c$ and its corresponding projection operator $P_c = I - P_d$ also reduces A. Thus the operator A is reduced to two parts, one operating in $\mathcal{H}_d$ and the other in $\mathcal{H}_c$.

Dirac assumes a hybrid position. He develops his theory as if he were working only with operators in $\mathcal{H}_d$. Yet he uses it later on for operators in $\mathcal{H}_c$ for which that theory is not applicable. The resulting inaccuracies in the mathematical language have caused most of the problems in Dirac's formalism.

The solution for this problem is to adopt from the beginning a formalism which is valid for both types of spectra and this can be actually done with relative ease.

The starting point is the definition of the spectrum given in the introduction as an example. As an illustration let us prove for instance theorem (i) in paragraph 9 on page 31. We can state it in the following somewhat sharper form:

Theorem: The spectrum of a self-adjoint operator is a closed subset of the real line.

Proof: If $z = x + iy$, $y \neq 0$ then $g = (A - z)f$ has a norm bounded below by

$$||g||^2 = ||(A - x)f||^2 + y^2||f||^2 \geq y^2||f||^2.$$

Hence $R_z = (A - z)^{-1}$ exists and is bounded by

$$||R_z||^2 \leq 1/y^2$$

therefore $z \in \rho(A)$.

In order to show that Λ is closed it suffices to show that $\rho(A)$ is open and this means around every $z \in \rho(A)$ one can construct a small circular disk entirely contained in $\rho(A)$. We shall omit details.

The discrete part of the operator A (that is its reduction to $\mathcal{H}_d$) behaves in many respects like a self-adjoint operator in a finite dimensional space. In particular one has an expansion theorem. Indeed the normalized eigenvectors ϕ_n ($n = 1, 2, \ldots$) form a complete orthonormal system in $\mathcal{H}_d$, so that any vector $f \in \mathcal{H}_d$ can be developed

$$f = \sum_{n=1}^{\infty} x_n \phi_n \quad x_n = (f, \phi_n). \tag{18}$$

This is sometimes called the *expansion theorem*. Mathematically it is a special case of the spectral theorem.

Although the operator A in the discrete subspace $\mathcal{H}_d$ has many features in common with the finite dimensional operators there are some differences which must be kept in mind.

The first of these is the fact that the sum in (18) may be an infinite sum. This means there is a problem of convergence and convergence presupposes a topology. The topology in which the sum (18) converges is the so-called *strong* topology. A sequence of vectors f_n is said to converge to f in the strong topology (written $f_n \rightarrow f$) if

$$\|f_n - f\| \rightarrow 0 \quad \text{for } n \rightarrow \infty. \tag{19}$$

An important property of Hilbert space is the fact that it is complete. This means any sequence of vectors f_n which satisfies the Cauchy condition, that is for any $\epsilon > 0$ there exists an N such that

$$\|f_n - f_m\| < \epsilon \quad \text{if} \quad n > N, m > N,$$

has the property that it converges to a unique $f \in \mathcal{H}$. The strong topology induced by the norm is contrasted with the *weak* topology induced by the scalar product. A sequence f_n converges weakly to a limit f, and we write $f_n \rightarrow f$ if for every $g \in \mathcal{H}$,

$$(f_n, g) \rightarrow (f, g). \tag{20}$$

This topology is called weak because strong convergence implies weak convergence but the converse is only true in finite dimensional spaces. The standard example to show this is an infinite orthonormal system $\{\phi_n\}$. Any such system converges weakly to zero but does not converge strongly at all since it is not even a Cauchy sequence. In formulae

$$(\phi_n, f) \rightarrow 0 \quad \forall f \in \mathcal{H}$$

but

$$\|\phi_n - \phi_m\| = \sqrt{2} \quad \text{for} \quad n \neq m.$$

The problem of convergence of such infinite sums such as (18) (or generalizations of them in the form of integrals) is the main difficulty which one has to face if one wants to take seriously Dirac's suggestion on page 40 to extend the Hilbert space to a larger space. So far there is no simple general solution for this problem known, or rather there are several possibilities for solving it since a Hilbert space can be embedded into a larger space in many different ways and one has to choose one of them in a rather arbitrary manner. It is for this reason that we find this suggestion impractical.

A second remark concerning the infinite discrete case refers to its spectrum. To be sure Λ_d is a denumerable subset of the real line. But such a subset may have accumulation points which may or may not belong to Λ_d. Since the spectrum is defined as a closed set we have in the discrete case $\overline{\Lambda} = \Lambda_d$, where the bar signifies the closure of the set Λ_d in the usual topology of the reals. In the infinite case it may very well be that $\Lambda_d \subsetneq \overline{\Lambda}$.

In order to show that we are not discussing any pathological situations let us exhibit a simple example where $\overline{\Lambda}_d = \Lambda = [0, 1]$ the closed interval from 0 to 1 for a bounded self-adjoint operator A with only discrete eigenvalues.

To this end we denote by $\{\phi_n\}$ $(n = 1, 2, \ldots)$ a complete orthonormal system and by r_n an enumeration of the rational numbers between 0 and 1. The operator A is then defined by

$$A\phi_n = r_n\phi_n$$

and extended to all of $\mathcal{H}$ by linearity. It is self-adjoint and bounded and its continuous spectrum is zero. Yet $\Lambda = [0, 1]$ and $\Lambda_d \equiv \{r_n\} \subset \Lambda$.

The rest of section 9 is devoted to special corollaries of the spectral theorem for self-adjoint operators which we shall discuss in the next section.

5. THE SPECTRAL THEOREM

The spectral theorem for self-adjoint operators in Hilbert space is a typical example of the general strategy sketched in the introduction of transferring properties and theorems from the finite to the infinite dimensional case. In finite dimensions the set of normalized vectors ϕ_n which satisfy equation (1) are a complete orthonormal system. The spectral theorem is the appropriate generalization of this fact for operators in infinite-dimensional space.

In order to show this we reformulate first this result for the finite-dimensional case in such a way that the new formulation is completely equivalent to the above statement and such that it can be transferred without change to the infinite dimensional case. In this manner we arrive at the spectral theorem.

Let us introduce the projection P_n with range ϕ_n, so that

$$P_n\phi_m = \delta_{nm}\phi_n.$$

Denote by Δ any Borel subset of the real line, for instance an interval and define

$$E_\Delta = \sum_{\lambda_n \in \Delta} P_n. \tag{21}$$

Since the P_n are all orthogonal to each other $E\Delta$ is again a projection. In particular if $\Delta = (-\infty, \lambda]$ we denote the corresponding E_Δ by

$$E_\lambda = E_{(-\infty, \lambda]}. \tag{22}$$

Every projection defines a subspace E_Δ and in the following we shall speak about subspaces and projections interchangeably. The family of projections are said to define a *spectral measure*, because of the following obvious properties:

(1) $E_\phi = 0 \quad E_R = I$

(2) if Δ_i $(i = 1, 2, \ldots)$ are disjoint $\Delta_i \perp \Delta_k$ for $i \neq k$ then

$$E_{\cup \Delta_i} = \Sigma E_{\Delta_i}.$$

This spectral measure is associated with A and in fact uniquely determined by it through the following formula:

Let $f \in \mathscr{H}$ then

$$(f, Af) = \int_{-\infty}^{+\infty} \lambda d(f, E_\lambda f). \tag{23}$$

This integral is to be understood as a Stieltjes integral since the non-decreasing function $(f, E_\lambda f)$ may have jumps at the eigenvalues of the operator A.

This last integral is sometimes written shorter without reference to a particular vector f in the form

$$A = \int_{-\infty}^{+\infty} \lambda dE_\lambda. \tag{24}$$

It means exactly the same thing as (23).

The spectral theorem gives a most convenient tool for developing the functional calculus for operators. For instance if $u(\lambda)$ is a function (whose only property needed is measurability with respect to the spectral measure) then we can define an operator

$$u(A) = \int_{-\infty}^{+\infty} u(\lambda)dE_\lambda. \tag{25}$$

In particular if u is some power of λ then we obtain for A^m ($m=2$, 3, . . .)

$$A^m = \int_{-\infty}^{+\infty} \lambda^m dE_\lambda.$$

We can leave it to the reader to verify that all the theorems stated and proved in the rest of paragraph 9 are simple corollaries of the spectral theorem.

6. OBSERVABLES

Paragraph 10, page 34, devoted to observables, is particularly interesting and important because here the mathematical formalism so far developed is combined with its physical interpretation.

The statements referring to the physical interpretation can all be reworded slightly to make them correct and sufficiently general also from the mathematical point of view.

Let us illustrate this with a few examples. The italicized physical interpretations concerning eigenstates of physical observables could for instance be stated as follows:

If the dynamical system is in a state belonging to the subspace E_Δ associated with a real dynamical variable A, then a measurement of A will certainly give a result contained in the set Δ.

Conversely, *if the system is in a state such that a measurement of a real dynamical variable A is certain to give a result contained in the subset $\Delta \subset R$ then the state is contained in E_Δ.* Similarly we can modify the corresponding statement of page 36.

Any result of a measurement of a real dynamical variable is contained in its spectrum. Conversely, any subset of the spectrum may contain the result of some measurement of the dynamical variable for some state of the system.

This part of the physical interpretation amounts essentially to a simple translation of one language into another without essential change of content. However there are some statements in this section which go considerably beyond the empirical content of quantum physics. We shall mention the following:

(1) At the beginning of this section Dirac explains that the result of a measurement must always be a real number and he uses this to justify that observables must be represented by self-adjoint operators.

He briefly mentions the possibility of complex dynamical variables and dismisses it as inadmissible because the real and imaginary part may not commute.

The conclusion drawn from this is incorrect since there exists a large class of complex operators with the property that their real and imaginary parts do commute. In that case these two parts may be measured simultaneously without one disturbing the other.

The operators which have this property are called normal operators by the mathematician. An operator N is normal if and only if it commutes with its adjoint

$$NN^* = N^*N. \tag{26}$$

(2) On page 36 Dirac stipulates that if a measurement of a quantity which gave the result in a subset Δ then after the measurement the system is in a state contained in E_Δ.

This is not always the case. This point was first discussed by Pauli,[5] who gave examples of measurements where this property is not verified. He introduced therefore the distinction of measurements of the first kind and measurements of the second kind, the first kind being those for which the above property is satisfied.

Thus the last sentence of the first paragraph on page 36 must be qualified with the remark that it is only true for measurements of the first kind.

(3) I have found no consistent interpretation of the last paragraph beginning on page 36. Dirac says: 'Not every real dynamical variable has sufficient eigenstates to form a complete set.'

From the context we have interpreted the notion of dynamical variable to be represented in the theory by a self-adjoint operator. The spectral theorem quoted in the preceding section is in contradiction with Dirac's statement quoted in the preceding paragraph. The conclusion could only be that real dynamical variables are something else than self-adjoint operators.

The only other candidates known to me would be symmetric operators which are not self-adjoint. Is this the meaning of this cryptic statement?

(4) On page 37 Dirac asks the question whether every observable can be measured. In other words he asks whether every self-adjoint operator represents a measurable quantity. He answers yes in theory but adds that in practice it may be very difficult to devise an apparatus which would measure some particular observable (self-adjoint operator).

We believe in this generality the answer is incorrect. There are systems which have superselection rules. In those systems the Hilbert space is split up into two or several subspaces and no self-adjoint operator with matrix elements connecting these subspaces can be measured. A typical example of this kind is the nucleon system containing the proton and the neutron as subsystems. A self-adjoint operator connecting these two subsystems would have for eigenstate a coherent superposition of a neutron and a proton state and no such states exist in the real world.

Superselection rules are related to 'supersymmetries' (cf. ref. 6). The particular case that we have been discussing here is intimately connected with the gauge invariance of the electromagnetic interaction. In fact as was pointed out in the mentioned reference: gauge invariance implies superselection rules with respect to the total charge quantum number.

7. COMMUTABILITY AND COMPATIBILITY

Paragraph 13, with this title, ends with the following conclusion: *From the point of view of general theory, any two or more commuting observables may be counted as a single observable, the result of a measurement of which consists of two or more numbers.*

Expressed mathematically this would mean the following: Given two self-adjoint operators A_1 and A_2 which commute, then there exists a third self-adjoint operator X and two functions u_1 and u_2 of a real variable such that the two operators can be expressed in the sense of the functional calculus

$$A_1 = u_1(X)$$
$$A_2 = u_2(X).$$

(27)

This theorem is entirely correct although the proof given by Dirac is not.

Some precaution is necessary if the operators A_1 and A_2 are unbounded, since commutability of unbounded operators is a non-trivial matter because of domain problems. However, for the sake of illustrating the essential content of the theorem, it is sufficient to discuss bounded operators only.

As usual with this kind of theorem, it is easier to formulate them and grasp their significance if one generalizes a bit. Instead of looking at just one pair of a finite set of mutually commuting operators we might as well look at an entire set of operators which commute with one another.

Let γ be any set of self-adjoint, bounded, and pairwise commuting operators. An easy way to construct all the operators which commute with all the γ and with each other is the following procedure due to von Neumann.

Let us denote by $\gamma' = \{T | TA = AT \;\; A \in \gamma \text{ and } \|T\| < \infty\}$, in words: γ' is the set of all bounded operators T which commute with all the operators in γ. This set γ' is not the set we want but it is an intermediary one needed for the construction of the desired set. The latter is obtained by repeating the same operation once more which leads to γ'' defined as the set of all bounded operators which commute with all operators in γ'.

The set $\gamma'' = \mathscr{A}$ is an algebra, that is it contains with any pair T_1, T_2 also the sum $T_1 + T_2$ and the product $T_1 . T_2$. Furthermore, it has the *-property: If $T \in \mathscr{A}$, then $T^* \in \mathscr{A}$. It is also the largest set of bounded operators which commute with all the operators in γ.

It is easily seen from the construction that the algebra $\mathscr{A}$ is abelian, a property which may be expressed in two different ways, either by saying $T_1 T_2 = T_2 T_1 \;\; \forall T_1$, $T_2 \in \mathscr{A}$ or more elegantly by the inclusion relation

$$\mathscr{A} \subseteq \mathscr{A}'. \tag{28}$$

In the special case that $\mathscr{A} = \mathscr{A}'$, we say that $\mathscr{A}$ is *maximal abelian*, and the self-adjoint operators $A_i \in \gamma$ of the set γ are said to be a complete set of *commuting observables*.

The following is a restatement of the main content of paragraph 13 in Dirac's book:

Theorem (von Neumann): Given an abelian von Neumann algebra $\mathscr{A}$ of bounded operators, there exists a bounded self-adjoint operator $X \in \mathscr{A}$ and functions $u(\lambda)$ ($\lambda \in \mathbf{R}$) measurable with respect to the spectral measure of X such that every $T \in \mathscr{A}$ can be expressed as a function

$$T = u(X).$$

Note that the function $u(\lambda)$ need only be measurable and it can be much more complicated than an analytical, differentiable or even merely continuous function. That functions of such generality must be admitted in order to obtain a simple theorem is seen from the following elementary example taken from physics:

Suppose we consider a particle moving in a plane. The two position coordinates Q_1, Q_2 are two commuting, self-adjoint operators. According to the Dirac–von Neumann theorem, there must exist a self-adjoint operator X and two functions $u_1(\lambda)$ and $u_2(\lambda)$ such that

$$\begin{aligned} Q_1 &= u_1(X) \\ Q_2 &= u_2(X). \end{aligned} \tag{29}$$

The difficulty starts if one tries to construct such a function. It is an incredibly complicated affair and cannot be done explicitly in any simple manner.

The upshot of all this is that one becomes rather convinced that such a construction cannot possibly have any physical significance. This casts doubt on the physical relevance of the theorem as well as Dirac's contention that every self-adjoint operator should represent an observable even in theory only.

8. THE SPECTRAL REPRESENTATION
(DISCRETE CASE)

Chapter III in Dirac's book is entitled 'Representations', and in its introduction he justifies the use of representations of the abstract quantities by the use of numbers in order to simplify the work for the application of the theory to particular problems. The procedure is compared with the use of coordinate axes in geometry.

Just as it is in geometry so the representation of the abstract quantities by numbers is not unique and one of the chief problems for the practical working physicist is to choose representatives that are as simple as possible, and that are adapted to the special conditions of the problem.

The tactics to be used in order to achieve this end is to choose basic bra-vectors which are the eigenbras of a set of commuting observables. Since such objects do in general not exist in Hilbert space, we have to modify these tactics in a similar way as we had to do for the definition of the spectrum.

We shall explain the procedure again first in the simplest finite-dimensional case and then generalize it in a number of steps to the infinite and degenerate case.

Let A be a self-adjoint operator in a space of dimension $n < \infty$, and assume that the eigenvalues λ_i $(i = 1, 2, \ldots, n)$ are all non-degenerate. This means the system of equations

$$A\phi_i = \lambda_i\phi_i \quad (i = 1, 2, \ldots, n)$$

has for each i only one linearly independent solution and $\lambda_i \neq \lambda_k$ for $i \neq k$. We shall agree to call a spectrum $\{\lambda_i\} = \Lambda$ with this property a *simple spectrum*.

The first problem is to transcribe the notion of *simple spectrum* into a different but equivalent form which does not make use of the notion of eigenfunctions (eigenkets or eigenbras) since such objects do not always exist in the infinite-dimensional Hilbert space.

We can get a heuristic feeling of what such a new formulation would look like if we examine the extreme degenerate case, that is all λ_i are equal to λ. In that case the operator A is simply a multiple of the unit operator $A = \lambda . I$. Now such an operator is characterized by the fact that everything commutes with it, so that $\{A\}'$ consists of the set of *all* bounded operators.

Conversely if the eigenvalues are all different from each other then we expect the $\{A\}'$ to be minimal and hence the algebra $\mathscr{A} \equiv \{A\}''$ to be maximal. Thus we conjecture the following theorem.

Theorem: The necessary and sufficient condition that the spectrum of the self-adjoint operator A is simple is that the von Neumann algebra $\mathscr{A}$ generated by it is maximal abelian

$$\{A\}'' \equiv \mathscr{A} = \mathscr{A}'. \tag{30}$$

This theorem is actually true and its proof is quite easy in the finite dimensional case.

It is clear that this theorem does go in the right direction since it replaces one definition of the simple spectrum (which cannot be used in Hilbert space) by another equivalent one which can be used. Before we proceed further with it we introduce a second equivalent form which turns out to be quite useful for carrying out calculations. It is based on the notion of the *cyclic vector*. But first again some heuristic remarks:

The unit operator can also be characterized by the fact that it leaves every vector fixed, while an operator with simple spectrum moves them around in space in a maximal manner. In order to express this property it is useful to have a measure of the *degree of mobility* for an operator.

Evidently not all vectors are moved around by the operator to the same degree. For instance the eigenvectors of A are only multiplied with scalars. It is therefore natural to suspect that the vectors which have the highest degree of mobility are the superpositions of all the eigenvectors. Let us call such a vector a *cyclic vector* and denote it by g:

$$g = \sum_{r=1}^{n} x_r \phi_r \quad x_r \neq 0 \quad (r = 1, 2, \ldots, n). \tag{31}$$

We then have the following theorem:

Theorem 2: Let A be any self-adjoint operator in a finite dimensional space $\mathscr{H}$, g a cyclic vector then any vector $f \in \mathscr{H}$ can be written in the form

$$f = Tg \quad \text{with} \quad T \in \{A\}'. \tag{32}$$

We shall omit the proof, since again it is quite easy.

We shall call such a g also a *cyclic vector* with respect to the algebra $\mathscr{A}' = \{A\}'$.

Combining the last two theorems we obtain a new characterization of the simple spectrum in the form:

Theorem 3: The operator A has a simple spectrum if and only if the algebra $\mathscr{A} = \{A\}''$ admits a cyclic vector. Furthermore in that case the representation

$$f = Tg \quad T \in \mathscr{A} \tag{33}$$

is unique.

It follows from the previous remarks on von Neumann algebras and this last theorem that to every $f \in \mathscr{H}$ we can associate a function $u(\lambda)$ (which in the case of finite dimensions is even a polynomial of degree at most n) in such a way that

$$f = u(A)g.$$

We obtain a representation of f as an element in a function space if we make the correspondence into an isometry. This means we want to define in the linear space of functions $u(\lambda)$ a norm such that it is equal to the norm of the corresponding vector.

This is quite easy. It suffices to calculate

$$||f||^2 = || \int u(\lambda) dE_\lambda g ||^2 = \int |u(\lambda)|^2 d(g, E_\lambda g). \tag{34}$$

From this formula we see the following:

The space $L^2(\mathbf{R}, \mu)$ consisting of complex-valued functions $u(\lambda)$ ($\lambda \in \mathbf{R}$) endowed with the norm

$$||u||^2 = \int |u(\lambda)|^2 d\mu,$$

where μ is the measure generated by the non-decreasing function $(g, E_\lambda g)$, is isomorphic with the Hilbert space $\mathscr{H}$. Furthermore in this isomorphism the operator A appears as a multiplication operator since

$$Af = Au(A)g \leftrightarrow \lambda u(\lambda).$$

We should mention here that the objects of the space $L^2(\mathbf{R}, \mu)$ are not individual functions $u(\lambda)$ subject to the norm condition

$$\int |u(\lambda)|^2 d\mu < \infty,$$

but rather equivalence classes of such functions with respect to the measure μ.

Two functions $u_1(\lambda)$ and $u_2(\lambda)$ are in the same equivalence class if

$$\int |u_1(\lambda) - u_2(\lambda)|^2 d\mu = 0.$$

The correspondence mentioned above is one–one not with individual functions but only with equivalence classes of such functions.

This point can often be suppressed since we can always pick out representatives from the equivalence classes but sometimes it is essential to be aware of it.

Because of this remark, we note that in the special example that we have discussed only the values $u(\lambda_k) = x_k$ of the functions $u(\lambda)$ are relevant. Hence the correspondence $f \longleftrightarrow u(\lambda)$ is one–one with polynomials of degree at most n.

We can avoid this ambiguity in this special case by considering the space of finite sequences $x = \{x_1, \ldots, x_n\}$ endowed with the norm

$$||f||^2 = ||x||^2 = \sum_{k=1}^{n} |x_k|^2 |(\phi_k, g)|^2. \tag{35}$$

Note that this is exactly the same expression as (34) only written as a sum instead of a Stieltjes integral. The quantities x_k can be expressed as a scalar product

$$(\phi_k, g) x_k = (\phi_k, f),$$

and (35) is nothing else than the relation

$$||f||^2 = \sum_{k=1}^{n} (f, \phi_k)(\phi_k, f).$$

This method of obtaining a representation of the abstract vector $f \in \mathcal{H}$ in terms of numbers $\{x_k\}$ or equivalence classes of functions $u(\lambda)$ suffers still from the slight defect that it depends on the arbitrary choice of the cyclic vector g. We can of course make a special choice, dictated by simplicity, for instance

$$g = \sum_{k=1}^{n} \phi_k.$$

In that case all the above formulae simplify in such a way that the vector g disappears.

We are now ready to compare the representation of abstract vectors $f \in \mathcal{H}$ used by Dirac with the one outlined above, which, we repeat, is completely equivalent to it for the particular case under discussion.

Method 1 (Dirac): Let A be a self-adjoint operator representing a complete set of commuting observables in a finite dimensional space.

Every eigenvalue λ_k of A determines a basic bra $<\phi_k|$. A general ket f is represented by the bracket

$$\langle\phi_k|f\rangle = x_k. \tag{$36)_1$}$$

This correspondence of f with the set $\{x_k\}$ is linear and such that

$$||f||^2 = \sum_{k=1}^{n} |x_k|^2$$

and

$$\langle\phi_k|Af\rangle = \lambda_k x_k. \tag{$37)_1$}$$

Method 2: Let A be as before. It generates a maximal abelian von Neumann algebra $\mathscr{A} = \{A\}''$ and therefore admits cyclic vectors. Let

$$g = \sum_{k=1}^{n} \phi_k$$

be one of them then every $f \in \mathscr{H}$ determines a unique set of numbers $f \longleftrightarrow \{x_k\}$ such that

$$f = u(A)g \longleftrightarrow \{x_k\}, \quad \text{where } u(\lambda_k) = x_k. \tag{$36)_2$}$$

Furthermore the correspondence is linear and such that

$$||f||^2 = \sum_{k=1}^{n} |x_k|^2$$
$$Af \longleftrightarrow \{\lambda_k x_k\}. \tag{$37)_2$}$$

A comparison of the two methods shows that the essential difference is in the manner of establishing explicitly the correspondence between the abstract vectors f on the one hand and their representatives $\{x_k\}$ on the other.

In the first method this correspondence is furnished by the bracket between the ket f and the eigenbras of the operator A (formula $(36)_1$).

In the second method it is established by the correspondence f with functions $u(\lambda)$ ($\lambda \in \mathbb{R}$) which have fixed values x_k at the points λ_k (formula $(36)_2$).

The basic properties of the representatives which are all that is needed for actual calculations are contained in formulae $(37)_1$ and $(37)_2$ and they are essentially the same for the two methods.

Method 1 cannot be generalized to Hilbert space because basic bras do not exist for the continuous spectrum. Hence equation $(36)_1$ is meaningless in that case and it cannot be used for constructing a representative.

Method 2 on the other hand can be so generalized since it uses only concepts and theorems which are valid in Hilbert space also, even if the spectrum of A is continuous.

9. THE SPECTRAL REPRESENTATION
(SIMPLE SPECTRUM IN HILBERT SPACE)

The spectral representation for a self-adjoint operator A with simple spectrum in Hilbert space can be established with no effort after the preparation of the preceding section. It is of course method 2 which will be used. The main result of this method can be taken over with only minor changes. First we shall give the result and then we shall make some explanatory remarks.

We are given a self-adjoint operator A with simple spectrum in a Hilbert space $\mathcal{H}$. We want to establish an isometry of $\mathcal{H}$ with a function space $L^2(\mathbb{R}, \mu)$ consisting of (equivalence classes of) functions $u(\lambda)$ measurable with respect to the measure μ and normalized according to

$$||u||^2 = \int |u(\lambda)|^2 d\mu.$$

The correspondence is such that if

$$f \leftrightarrow \{u(\lambda)\}$$

then

$$Af \leftrightarrow \{\lambda u(\lambda)\}. \tag{38}$$

This is always possible. It is established in the following manner:

Since A has simple spectrum, the algebra $\mathcal{A} = \{A\}''$ is maximal abelian and therefore admits cyclic vectors. Let g be one of them and E_λ the spectral family of A. Then there exists a dense linear manifold denoted by $\{\mathcal{A}g\}$ consisting of all vectors $f = Tg$, $T \in \mathcal{A}$, with $T = u(A)$ for some $u(\lambda)$.

The correspondence (38) is then first established for all vectors in $\{\mathcal{A}g\}$ and it is then extended by continuity to all of $\mathcal{H}$. Furthermore calculation of the norm of $f = u(A)g$ yields

$$||f||^2 = \int |u(\lambda)|^2 d(g, E_\lambda g), \tag{39}$$

so that the measure $\mu(\lambda)$ is identified with

$$\mu(\lambda) = (g, E_\lambda g). \tag{40}$$

REMARKS

(1) The main difference between the finite and the infinite-dimensional case is the fact that in general the set $\{\mathcal{A}g\}$ is only a dense linear manifold and not the entire space $\mathcal{H}$. This is a technical detail which must be taken into account in the detailed mathematical discussion but it has no effect on the result.

(2) The measure μ depends on the choice of the cyclic vector. However all the measures are equivalent in the following precise sense:
If g_1 and g_2 are two different cyclic vectors then the two corresponding measures μ_1 and μ_2 have the same sets of measure zero. Two measures with this property are said to be *equivalent*. Thus the operator determines not a particular measure but only an *equivalence class* of measures.

(3) All the measures obtained with cyclic vectors are so-called finite measures. They satisfy

$$||g||^2 = \int d\mu < \infty.$$

An equivalence class may also contain infinite measures (for instance Lebesgue measure on $\mathbf{R}$).

One can pass from one μ to another equivalent one by a general formula involving the Radon–Nikodym derivative $d\mu/d\nu$ of one measure with respect to another.

Thus $\int d\mu = \int (d\mu/d\nu)d\nu$, where $d\mu/d\nu$ is a bounded positive function of the integration variable $\lambda \in \mathbf{R}$.

The passage from one measure to another corresponds in Dirac's method to a change of normalization of the eigenbras. In particular the so-called δ-function normalization implies that the measure adopted is Lebesgue measure.

(4) The preceding remark raises a difficult mathematical question: given an operator with simple continuous spectrum (that is without point eigenvalues), is it always true that the measure generated by $\mu(\lambda) = (g, E_\lambda g)$ is equivalent to Lebesgue measure on the spectrum Λ? A measure with this property is said to be absolutely continuous with respect to Lebesgue measure, and the function $\mu(\lambda)$ is then said to be absolutely continuous.

Dirac's formalism presupposes that this is always the case, but this is not true. It is one of the deeper results of real analysis that there exist functions which are continuous and non-decreasing but not absolutely continuous. For such functions one may very well have $\mu(\lambda_1) < \mu(\lambda_2)$ for $\lambda_1 < \lambda_2$, but $d\mu/d\lambda = 0$ except on a set of Lebesgue measure zero. Such functions are called singularly continuous.

The set of vectors $f \in \mathcal{H}_c$, such that $(f, E_\lambda f)$ is absolutely continuous, generate a subspace $\mathcal{H}_{ac} \subset \mathcal{H}_c$ which also reduces the operator A. The orthogonal complement of $\mathcal{H}_{ac}$ in $\mathcal{H}_c$ is called the singularly continuous subspace and it is denoted by $\mathcal{H}_{sc}$. Thus for any operator with simple spectrum we have the decomposition into *three* reducing orthogonal subspaces:

$$\mathcal{H} = \mathcal{H}_d \oplus \mathcal{H}_{ac} \oplus \mathcal{H}_{sc}.$$

There is no physically meaningful observable known which has singularly continuous spectrum. In view of Dirac's conviction that essentially every self-adjoint operator should represent, at least in theory, an observable it might be particularly interesting to study the reason for the non-occurrence of singularly continuous spectra in physical observables.

10. THE SPECTRAL REPRESENTATION
(GENERAL CASE)

When an observable A does not have a simple spectrum there are two possibilities of generalizing the results of the preceding section. Both of them are used in physical applications.

The first one follows closely Dirac's method of paragraph 14. The observable A is taken as the first one of a sequence of observables

$$\{A = A_1, A_2, \ldots, A_n, \ldots\} = \gamma,$$

finite or countably infinite in number which together form a complete set of commuting observables. As we have shown before this means $\gamma'' = \gamma'$.

Let
$$\lambda = \{\lambda_1, \lambda_2, \ldots\} \in \Lambda_1 \times \Lambda_2 \times \ldots \equiv \Lambda$$

be an element of the Castesian product of the spectra Λ_r of A_r. The spectral representation theorem then assert the existence of a measure μ (or rather a measure class as explained above), a space $L^2(\Lambda, \mu)$ and an isometric mapping of the space $\mathscr{H} \longleftrightarrow L^2(\Lambda, \mu)$ such that we have

$$\phi \longleftrightarrow u(\lambda)$$
$$A_r f \longleftrightarrow \lambda_r u(\lambda). \tag{41}$$

Such a theorem can and has been established but only under an additional hypothesis which guarantees that the operators $A_1, A_2, \ldots$ are in some sense independent of one another.[6] This hypothesis is not always easy to verify. In a practical case, however, there is usually no question that this condition is satisfied.

The second method of using a representation is not treated in Dirac's book. However, it is implicitly used in chapter VIII on collision problems where it is closely related to the variable α which appears from paragraph 49 on.

This method is based on the notion of the direct integral of Hilbert spaces. We shall give here a brief exposition of this concept in the special case of uniform multiplicity of the spectrum, a term which will explain itself as we go on.

M

Let $f: \mathbf{R} \to \mathscr{H}_0$ be a function from $\mathbf{R}$ to a fixed Hilbert space $\mathscr{H}_0$, so that $\forall \lambda \in \mathbf{R}, f_\lambda \in \mathscr{H}_0$. We denote the scalar product in $\mathscr{H}_0$ by $(\cdot, \cdot)_0$. If for any fixed $g \in \mathscr{H}_0$ the function $(g, f_\lambda)_0$ is measurable with respect to some measure μ we shall say f_λ is a μ-measurable function with values in $\mathscr{H}_0$. We shall only consider such functions in the following.

The set of functions $f = \{f_\lambda\}$ with the property

$$\|f\|^2 \equiv \int \|f_\lambda\|_0^2 d\mu(\lambda) < \infty$$

forms a Hilbert space under component-wise addition and scalar multiplication. We denote it by $L^2(\mathscr{H}_0, \mu) = \mathscr{H}$.

It is clear that the elements in $L^2(\mathscr{H}_0, \mu)$ determine the functions f_λ only up to sets of measure zero. We shall write in this case a.e.(μ) (read: almost everywhere with respect to the measure μ).

Consider now the function X from $\mathbf{R}$ to bounded operators in $\mathscr{H}_0$. We assume $X_\lambda f_\lambda$ to be measurable. Then the operation

$$\{f_\lambda\} \equiv f \to g = \{X_\lambda f_\lambda\} = Xf \quad \text{a.e.}(\mu)$$

is a linear operator in $\mathscr{H}$ which we shall call the direct integral of the X_λ. The following facts are not hard to verify:

(1) X is bounded iff the X_λ are essentially bounded;
(2) X is self-adjoint or unitary iff the X_λ are a.e.(μ);
(3) an operator B which commutes with all X is of the form $B = \{u_\lambda I_0\}$ a.e.(μ) where u_λ is a scalar valued function and I_0 is the unit operator in $\mathscr{H}_0$.

We shall call such operators X in $\mathscr{H}$ operators with *uniform multiplicity*. The dimension of $\mathscr{H}_0$ is the degree of multiplicity. Such operators occur very often in quantum theory.

A simple example is furnished by a free particle constrained to move in one direction only. Its energy in the non-relativistic case is given by $H = (1/2m)P^2$. The spectrum Λ of H is the positive real axis $\mathbf{R}^+$ and it has uniform multiplicity 2.

In elementary scattering theory the total Hamiltonian takes the form $H = H_0 + V$ where H_0 is the kinetic energy of a free particle and V is the interaction.

H_0 has usually the spectrum $\Lambda = \mathbf{R}^+$ and the multiplicity is uniform and ∞. The set of vectors $f_\lambda \in \mathscr{H}_0$ is then called the energy shell. It is well-known that the scattering operator S commutes with H_0 and it admits therefore a direct integral representation $\{S_\lambda\}$ where S_λ is a unitary operator a.e.(μ) on the energy-shell.

We can call the correspondence $f \leftrightarrow \{f_\lambda\}$ and $X \leftrightarrow \{X_\lambda\}$ a particular spectral representation with respect to the operator $A = \{\lambda.I_0\}$. Again we find that the operator A is diagonal in this representation in the sense that if

$$f \longleftrightarrow \{f_\lambda\},$$

then $Af \longleftrightarrow \{\lambda f_\lambda\}$.

11. REMARKS ON NOTATION

The notation in Dirac's mathematical formalism is a very important element. It is easy to learn and it is a great aid in practical calculations. That is why it has become widely adopted in the scientific literature.

It is therefore of great interest to preserve as much as possible the notational advantages of Dirac's formalism. This is possible if one reinterprets some of the symbolic bra-ket notation in a slightly different way.

We shall illustrate this with an example. We have not one but actually two objects to deal with. One refers to the scalar product in Hilbert space which we have denoted by (f, g) and we shall continue to do so. the other refers to the representatives of the abstract quantities $f, g, \ldots$ by functions in an L^2-space. We have used for such functions $u(\lambda)$ if they were complex-valued and f_λ if they were vector valued in a fixed space $\mathscr{H}_0$.

We shall now introduce for $u(\lambda)$ a new notation which combines the advantage of Dirac's bra-ket notation with a correct mathematical content. For the function $u(\lambda)$ which corresponds to the vector $f \in \mathscr{H}$ we shall write

$$u(\lambda) = \langle \lambda | f \rangle \in L^2(\lambda, \mu).$$

We have then the following operating rules

$$\langle \lambda | f \rangle + \langle \lambda | g \rangle = \langle \lambda | f+g \rangle$$
$$\langle \lambda | \alpha f \rangle = \alpha \langle \lambda | f \rangle$$
$$(f | \lambda) = \overline{\langle \lambda | f \rangle}$$
$$(f | g) = \int (f | \lambda) d\mu(\lambda) \langle \lambda | g \rangle.$$

Here $d\mu$ is the measure which may or may not be equivalent to Lebesgue measure on the set Λ.

Finally for $A_i \in \gamma$ we have

$$\langle \lambda | A_i f \rangle = \lambda_i \langle \lambda | f \rangle,$$

and if $v(A_1, A_2, \ldots)$ is some function of the observables

$$\langle \lambda | v(A_1, A_2, \ldots) f \rangle = v(\lambda_1, \lambda_2, \ldots) \langle \lambda | f \rangle.$$

These examples may suffice to show how Dirac's bra and ket notation, so effective as an operational device, may be based on a mathematically rigorous foundation.

12. REPRESENTATION OF OPERATORS

In paragraph 17 on page 69 we find the statement (together with four other equally questionable ones) *Any linear operator is represented by a matrix.*

The truth value of such a statement depends of course entirely on what one means with the word 'matrix'. The difficulty is not only with the continuous spectrum. Even if the spectral representation with respect to A is discrete and defines a complete orthonormal system $\{\phi_r\}$ we cannot claim that a given operator X is always represented by the matrix

$$X_{rs} = (\phi_r, X\phi_s).$$

The reason is, of course, that for unbounded operators the vectors ϕ_r need not be in the domain of definition D_X.

If the spectrum of A is continuous the difficulty is compounded since there are obvious counter examples, such as an integral operator plus a multiple of the unit operator. It is to meet these difficulties that Dirac has introduced his celebrated δ-function which he believes justifies the above statement. However this is not the case and the situation is in fact so complicated that it is much more economical to drop this mathematical fiction altogether and cope honestly with the reality of the situation.[9]

However, before doing this we should mention one remarkable exception where the above statement is in fact true. This is the case for the Hilbert space of coherent states. It is a space consisting of a certain class of analytical functions $f(z)$ with the property that the annihilation operator $a = (1/\sqrt{2})(Q + iP)$ is a differential operator $(af)(z) = df/dz$ and the creation operator is a multiplication operator

$$(a^*f)(z) = zf(z).$$

This property together with the fact that the two operators are Hermitian conjugates of each other determines the measure μ of the metric and with it the space. The set of oscillator eigenfunctions ϕ_n $(n = 0, 1, 2, \ldots)$ which satisfy

$$a^*a\phi_n = n\phi_n,$$

define in it a complete orthonormal system.

ON BRAS AND KETS   165

This space has the remarkable property that every linear operator X can be written as a bona fide integral operator, so that

$$(Xf)(z) = \int X(z, \xi) f(\xi) d\mu(\xi),$$

with a regular Kernel $X(z, \xi)$.

In particular for the identity operator one finds

$$f(z) = \int I(z, \xi) f(\xi) d\mu(\xi)$$

with a perfectly regular function $I(z, \xi)$ defined everywhere on $\mathbb{C} \times \mathbb{C}$. We might call $I(z, \xi)$ a *self-reproducing kernel*. It does exactly what Dirac's δ-function is supposed to do without, however, having any singularity whatsoever.[3]

This is a special case. We return now to the general case of the representation of an operator A in an $L^2(\Lambda, \rho)$ space as an integral operator. Here the situation is much more complicated. One of the difficulties is that this property is not unitarily invariant. This means that if A is such an integral operator, so that we may write

$$(Af)(\lambda) = \int A(\lambda, \mu) f(\mu) d\rho(\mu),$$

it is not true that the operator $UAU^{-1} = A'$ for unitary U is also such an operator.

It is thus meaningful to ask the question whether a given operator A is unitarily equivalent to an operator which has such an integral representation.

One of the first results obtained in this direction is due to von Neumann[7] who proved that the necessary and sufficient condition that a self-adjoint operator be unitarily equivalent to an integral operator of Carleman type is that 0 is a condensation point of the spectrum.

Later, in a sharpening of von Neumann's result, Misra, Speiser and Targonski[8] have shown that this is also true for normal operators. Futhermore everyone of a set of unitarily equivalent operators is such an integral operator if and only if one and hence all of them are of Hilbert–Schmidt type.[8]

Another line of research has recently been pursued by Kato.[4] He introduces the extremely useful concept of a *smooth operator* with respect to a self-adjoint operator H. In order to express this notion numerically Kato introduces a smoothness norm of an operator A by the following procedure.

Consider the unitary group $V_t = e^{-iHt}$ associated with H. For any unit vector ϕ we then consider $||AV_t\phi||^2$. It may or may not be true that this quantity tends to zero as $t \to \pm \infty$. (It never does if ϕ has a component in $\mathcal{H}_d$ and it always does if $\phi \in \mathcal{H}_{ac}$.)

If it does tend to zero then the speed with which it does is a measure of the 'smoothness' of the operator A with respect to H.

Thus one is lead to introduce the 'smoothness norm' with respect to H defined by

$$||A||^2{}_H = \sup_{\phi \in S} \frac{1}{2\pi} \int_{-\infty}^{+\infty} ||AV_t\phi||^2 dt. \tag{42}$$

If $||A||^2{}_H < \infty$ then the operator A is called H-smooth. Kato then proves a number of remarkable theorems, from which we select the following two which are relevant for our purpose.

Theorem 1: If H is bounded and A is H-smooth then A vanishes in $\mathcal{H}_d$ and in $\mathcal{H}_{sc}$.

Theorem 2: Let H be bounded with a simple absolutely continuous spectrum then the operator $B = A*A$ is an integral operator in the spectral representation of H with an essentially bounded kernel if and only if A is H-smooth.

In spite of its special nature theorem 2 is extremely useful in applications since it is just in the context of scattering theory that one encounters the smooth operators and this theory constitutes by far the most important application of the theory of continuous spectrum.

There is an obvious generalization of this result for operators with degenerate spectrum. (Kato actually proves a much more general theorem than the one we quote here.)

As a corollary we mention the following facts: If H has only absolutely continuous spectrum then every Hilbert–Schmidt operator A is H-smooth.

One application of this result is the following: Under relatively weak conditions on the iteraction operator V for the total Hamiltonian $H = H_0 + V$ in a scattering problem one can prove that the operator $S_\lambda - I \equiv R_\lambda$ on the energy shell is Hilbert–Schmidt. Hence it is an integral operator (we mean in the proper sense of the word, not in the extended sense of Dirac) on the energy shell, that is, in this case, on the unit sphere. The corresponding matrix elements $(\alpha|R_\lambda|\alpha_0)$ are the scattering amplitudes for the scattering from the direction α_0 into the direction α.

These and similar other theorems show that a large part of the formal manipulations of Dirac can be used also for operators. This is probably the chief reason why most physicists have not felt a stronger need for a more rigorous mathematical foundation of elementary quantum mechanics.

13. CONCLUDING REMARKS

The foregoing sections have, I hope, demonstrated my point: Dirac's beautiful and powerful, but mathematically ambiguous, formalism can be translated into a precise mathematical frame without obscuring the physical content of the theory. The necessary changes are not as profound as most physicists fear. Certain formulae and the vocabulary must be changed a little, all a matter of habits.

The new method retains many of the formal advantages of the old one. Yet many problems which were outside the possibility of rigorous treatment with the old formalism become accessible to such treatment in the new one. This is particularly the case for problems in scattering theory, relativistic or unrelativistic. Much progress has been made in this field thanks to the more powerful mathematical tools which can be brought to bear on these problems in the new formalism. It is to be hoped that in this manner the fruitfulness of Dirac's genius can be extended into domains where it would not have been possible before.

REFERENCES

1. P. A. M. Dirac, *Rev. Mod. Phys.* **34**, 592 (1962).
2. All quotations without specifications are taken from the fourth edition of Dirac's book: *The Principles of Quantum Mechanics* (Clarendon Press, Oxford: 1958).
3. This Hilbert space of analytical functions has been used particularly in connection with coherence problems in the quantum theory of radiation. It seems to have been rediscovered by many people independently. It is quite remarkable that Dirac himself was one of the first to introduce this space in a paper published in the *Ann. Inst. H. Poincaré*, **11**, 15 (1949). This paper seems to have been overlooked by all later workers on this problem. However, Dirac did not notice the above mentioned representation property for linear operators.
4. T. Kato, 'Smooth operators and commutators', *Studia Mathematica* XXXI, 535 (1968).
5. W. Pauli, *Handbuch der Physik*, vol. v, part 1, 1.
6. J. M. Jauch and B. Misra, *Helv. Phys. Acta* **38**, 30 (1965).
7. J. von Neumann, *Actualités Scient. et Ind.* **229**, Paris (1935).
8. B. Misra, D. Speiser and G. Targonski, *Helv. Phys. Acta* **36**, 963 (1963).
9. J. von Neumann, *J. f. Math.* **161**, 208 (1929).

IO

The Poisson Bracket

C. Lanczos

The 'Poisson bracket', which was destined to play such a decisive role in the higher echelons of analytical dynamics, made its first appearance in 1809, in a Mémoire of S. D. Poisson,[1] which dealt with the general perturbation problem of analytical mechanics. This paper was a direct sequel of an earlier investigation of Lagrange,[2] who was interested in the problem of astronomical perturbations, namely the problem of how the 'constants' (today we would say 'parameters') of the Keplerian ellipses are slowly changing under the influence of the mutual attraction of the planets. In the same year he investigated the general perturbation problem of analytical mechanics in a truly fundamental paper,[3] which introduced a certain bracket expression (the Lagrangian bracket), denoted by (a, b). This paper was the source of Poisson's far reaching investigations in the following year.

It seems of interest to observe that the French analysts operated already in phase space and obtained the most important consequences of Hamilton's canonical equations, much before Hamilton (and later Jacobi) appeared on the platform. Since the Lagrangian equations of motion could be written as a first order system for the momenta p_i:

$$\dot{p}_i = \frac{dp_i}{dt} = \frac{\partial L}{\partial q_i},\tag{1}$$

where the p_i are defined by

$$p_i = \frac{\partial L}{\partial \dot{q}_i}\tag{2}$$

it seemed reasonable to consider the q_i and p_i as the basic variables of the mechanical system, although this was still far from recognizing that the second equation is equivalent to

$$q_i = \frac{\partial H}{\partial p_i},\tag{3}$$

while the first equation could be put in the form

$$\dot{p}_i = -\frac{\partial H}{\partial q_i}, \tag{4}$$

which combined the two fundamental equations (1) and (2) into a unified system (3, 4) of remarkable simplity and beauty, replacing the original Lagrangian function L by the more fundamental Hamiltonian function $H(q_i, p_i, t)$; (ref. 4, pp. 166, 215).

The real significance of the Lagrange and Poisson brackets revealed itself only, when Jacobi investigated the general group of transformations ('canonical transformations') which left the 'canonical equations' (3, 4) invariant.[5] Poisson's fundamental theorem was that any bracket expression $[a, b]$ (using our present day nomenclature for Poisson's (a, b)), formed for any conjugate pair of variables of the perturbed system with respect to the unperturbed q_i, p_i, *remains constant with respect to time.*

The Poisson bracket $[u, v]$ is defined by the following operation:

$$[u, v] = \frac{\partial u}{\partial q_i}\frac{\partial v}{\partial p_i} - \frac{\partial v}{\partial q_i}\frac{\partial u}{\partial p_i}. \tag{5}$$

(Throughout this paper we will make use of Einstein's sum convention by automatically summing over two equal indices. In comparison the notations employed by Lagrange and Poisson are exceedingly clumsy. They wrote out explicitly every expression, without the use of indices, denoting a quantity like q_i by a, b, c, etc., which makes the reading of these papers excessively difficult.)

The bracket expression (5) possesses a number of magical properties. We see directly from the definition:

$$[u, v] = -[v, u], \tag{6}$$

and thus

$$[u, u] = 0, \tag{6a}$$

$$[u, v+w] = [u, v] + [u, w], \tag{7}$$

$$[\alpha u, v] = \alpha[u, v] + [\alpha, v]u, \tag{8}$$

$$\frac{\partial}{\partial a}[u, v] = \left[\frac{\partial u}{\partial a}, v\right] + \left[u, \frac{\partial v}{\partial a}\right], \tag{9}$$

$$[q_i, v] = \frac{\partial v}{\partial p_i}, \ [p_i, v] = -\frac{\partial v}{\partial q_i}. \tag{10}$$

Hence $\quad [w, [q_i, v]] - [[w, q_i], v] = \left[w, \dfrac{\partial v}{\partial p_i} \right] + \left[\dfrac{\partial w}{\partial p_i}, v \right]$

$$= \frac{\partial}{\partial p_i} [w, v] = [q_i, [w, v]], \tag{11}$$

and thus $\quad [w, [q_i, v]] + [q_i, [v, w]] + [v, [w, q_i]] = 0, \tag{12}$

or, denoting the operation of cyclic permutation by

$$[abc] = [a, [b, c]] + [b, [c, a]] + [c, [a, b]], \tag{13}$$

we obtain $\qquad\qquad\qquad [q_i vw] = 0. \tag{14}$

Of decisive importance was, however, Jacobi's profound extension of Hamilton's work by introducing the general group of transformations which left the Hamiltonian system (3, 4) invariant.[5] Since this system could be conceived as the result of a variational problem with the Lagrangian

$$L = p_i q_i - H(q_i, p_i), \tag{15}$$

the group of transformations which leaves the resulting variational equations invariant, demands merely the condition

$$p_i dq_i - P_i dQ_i = dS, \tag{16}$$

where dS is the total differential of a function $S(q_i, Q_i)$. This S can be eliminated by taking two independent differentiations d' and d'':

$$d''(p_i d'q_i - P_i d'Q_i) = d''d'S$$
$$d''p_i d'q_i - d''P_i d'Q_i + p_i d''d'q_i - P_i d''d'Q_i = d''d'S. \tag{17}$$

If we now exchange the sequence of the two differentiations and take the difference, the remaining terms can be put in the form

$$d''p_i d'q_i - d'p_i d''q_i = d''P_i d'Q_i - d'P_i d''Q_i. \tag{18}$$

This shows that the phase space possesses an invariant differential form in the sense of a *bilinear differential form*, associated with *two* independent infinitesimal directions (cf. ref. 8, p. 212), in remarkable contrast to the quadratic differential form ds^2 of a metrical space, which demands only *one* infinitesimal direction.

The functional relation between the old and new canonical coordinates q_i, p_i and Q_i, P_i can now be written down in the form of the following (necessary and sufficient) conditions:

$$(Q_i, Q_k) = 0, \quad (P_i, P_k) = 0, \quad (Q_i, P_k) = \delta_{ik}, \tag{19}$$

where we made use of the 'Lagrangian brackets', defined by

$$(a, b) = \frac{\partial q_i}{\partial a} \frac{\partial p_i}{\partial b} - \frac{\partial q_i}{\partial b} \frac{\partial p_i}{\partial a}. \tag{20}$$

Now let us assume that we have $2n$ independent functions $u_1, \ldots, u_{2n}$ of the $2n$ variables q_i and p_i. We can form the Lagrangian brackets (u_i, u_k) and likewise the Poisson brackets $[u_i, u_k]$. Then, if we write down the sum $(u_i, u_a) \cdot [u_k, u_a]$, making use of the rule of implicit differentiation, we obtain the following relation between the two types of brackets:

$$(u_i, u_a) \cdot [u_k, u_a] = \delta_{ik}. \tag{21}$$

This shows that the matrix of the Lagrange brackets and the matrix of the Poisson brackets are in an inverse transpose relation to each other; (cf. ref. 6, p. 300; ref. 8, p. 215; ref. 9, p. 152). Consequently the relations (19) can also be formulated in terms of the Poisson brackets:

$$[Q_i, Q_k] = 0, \quad [P_i, P_k] = 0, \quad [Q_i, P_k] = \delta_{ik}. \tag{22}$$

Now let us transform the q_i, p_i to *arbitrary* new variables Q_i, P_i. By the rule of implicit differentiation we obtain

$$[u, v]' = \left(\frac{\partial u}{\partial q_\alpha} \frac{\partial q_\alpha}{\partial Q_i} + \frac{\partial u}{\partial p_\alpha} \frac{\partial p_\alpha}{\partial Q_i} \right) \left(\frac{\partial v}{\partial q_\beta} \frac{\partial q_\beta}{\partial P_i} + \frac{\partial v}{\partial p_\beta} \frac{\partial p_\beta}{\partial P_i} \right)$$

$$- \left(\frac{\partial v}{\partial q_\alpha} \frac{\partial q_\alpha}{\partial Q_i} + \frac{\partial v}{\partial p_\alpha} \frac{\partial p_\alpha}{\partial Q_i} \right) \left(\frac{\partial u}{\partial q_\beta} \frac{\partial q_\beta}{\partial P_i} + \frac{\partial u}{\partial p_\beta} \frac{\partial p_\beta}{\partial P_i} \right)$$

$$= \frac{\partial u}{\partial q_\alpha} \frac{\partial v}{\partial q_\beta} [q_\alpha, q_\beta] + \frac{\partial u}{\partial p_\alpha} \frac{\partial u}{\partial p_\beta} [p_\alpha, p_\beta]$$

$$+ \left(\frac{\partial u}{\partial q_\alpha} \frac{\partial v}{\partial p_\beta} - \frac{\partial v}{\partial q_\alpha} \frac{\partial u}{\partial p_\beta} \right) [q_\alpha, p_\beta]. \tag{23}$$

If, however, the transformation is *canonical*, and thus the conditions (22) hold, then the right side is reduced to $[u, v]$ and we obtain

$$[u, v]' = [u, v]. \tag{24}$$

Any Poisson bracket is thus an *invariant of an arbitrary canonical transformation*.

Let us now consider the expression $[u \, v \, w]$, defined according to (13). Since it is composed of bracket expressions, it is an invariant of canonical transformations. By such a transformation we can choose u as one of our new variables, e.g. Q_1 (cf. ref. 8, p. 231). But then we see at once that the relation (14) can be generalized to

$$[u \, v \, w] = 0, \tag{25}$$

which is known as 'Jacobi's identity' (cf. ref. 7, p. 252; ref. 9, p. 151). We now understand the deeper reason for Poisson's Theorem concerning the time independence of the Poisson brackets in perturbation problems. The motion in time of a dynamical system can be conceived as a *succession of infinitesimal canonical transformations.* Such transformations cannot change the value of $[u, v]$ (cf. equation (24)), which means that $[u, v]$ must remain a *constant* throughout the motion.

The close relation of the Poisson brackets to the canonical equations has still another facet. Let us consider an arbitrary function $F(q_i, p_i)$ of the canonical variables q_i, p_i. Then

$$\dot{F} = \frac{\partial F}{\partial q_i} \dot{q}_i + \frac{\partial F}{\partial p_i} \dot{p}_i = \frac{\partial F}{\partial q_i} \frac{\partial H}{\partial p_i} - \frac{\partial F}{\partial p_i} \frac{\partial H}{\partial q_i} = [F, H]. \tag{26}$$

A particularly important application of this equation yields the *conservation of energy* theorem for conservative systems, whose Hamiltonian does not depend explicitly on t. Let us identify the previous F with $H(q_i, p_i)$, obtaining

$$\dot{H} = [H, H] = 0, \quad H = \text{const.} \tag{27}$$

We may add in parentheses that the canonical equations themselves can be put in the form

$$\dot{q}_i = [q_i, H],$$
$$\dot{p}_i = [p_i, H]. \tag{28}$$

The stream of evolution which leads from classical mechanics to the quantum mechanics or wave mechanics of our days, has at its fountainhead two great names: Hamilton and Jacobi. Hamilton's basic viewpoint was a method which was equally successful in mechanics as in geometrical optics. In both cases a certain integral had to be minimized: in mechanics the 'action', in optics the 'time'. In both cases a certain 'generating function', or 'Principal Function' gave the complete solution of the given problem. This generating function expressed the integral to be minimized as a function of $2n$ variables, namely the n position coordinates of the initial point and the end point of the (mechanical or optical) path. The generating function satisfied a certain partial differential equation with respect to the end coordinates and an analogous equation with respect to the initial coordinates; (cf. ref. 4, pp. 168, 631).

Jacobi took a much broader view concerning the 'generating function'. He studied the general group of *transformations* which left the canonical equations unchanged and characterized this group in terms of a generating function $S(q_i, Q_i)$ (cf. equation (16)). By choosing this function properly the canonical equations could be put in a trivially

simple form, which allowed their immediate integration. This demanded
only an arbitrary complete solution of one *single* partial differential
equation: the 'Hamilton–Jacobi differential equation'; (cf. ref. 8, p. 239).

When L. de Broglie discovered the 'matter waves',[10] Schrödinger,
who was familiar with the optico-mechanical analogy of Hamilton,
interpreted the Hamilton–Jacobi differential equation as the phase
equation of an optical field. Changing over to the amplitude equation
he established the famous 'Schrödinger wave-equation', which became
the corner-stone of 'wave mechanics'.[11]

Jacobi, however, was also the fountainhead of another stream of
thought, associated with the genius of P. A. M. Dirac. Jacobi discovered
the deeper significance of the Lagrange and Poisson brackets, as
corollaries of his transformation theory which played a fundamental
role in his dynamical theory.[5] Dirac was acquainted with the importance
of the Poisson brackets and endeavoured to base the canonical equations
entirely on these brackets, avoiding the explicit use of the partial
derivatives of H. He thought that the quantum phenomena indicated a
radical departure from the classical concepts, inasmuch as only *algebraic
operations* should take the place of derivatives. The derivatives with
respect to t should not be altered. But perhaps the Poisson bracket
should be reformulated in a completely novel way.

Heisenberg abolished the continuous Bohr orbits and replaced the
entire assembly of Bohr orbits by an infinity of discontinuous quantities,
characterized by two indices.[12] The 'product' (xy) of two such quantities
followed a composition scheme that Born recognized as the law of
matrix multiplication. Hence the transition from classical to quantum
mechanics had to occur by replacing the dynamical variables q_i, p_i of
classical mechanics by infinite matrices which followed a definite
commutation rule, derived from Heisenberg's product expression.[13]

Dirac's approach was different. He was aware of Heisenberg's
product rule and asked himself the following question (uninfluenced
by the work of Born, Heisenberg, Jordan, which appeared later[13]).
Suppose we formulate the canonical equations of Hamilton in terms
of the Poisson brackets; (cf. equations (26), (28)), but leaving the
definition of brackets free, except for demanding the following operational rules:

$$[u, v] = -[v, u],$$
$$[u, v+w] = [u, v] + [u, w],$$
$$\frac{\partial}{\partial a}[u, v] = \left[\frac{\partial u}{\partial a}, v\right] + \left[u, \frac{\partial v}{\partial a}\right].$$

(29)

How must we re-interpret the bracket expression $[u, v]$, in view of Heisenberg's product composition rule for quantum mechanical quantities? He proved that the re-definition of the Poisson bracket must occur according to the following law:

$$[u, v] = \frac{1}{i\hbar}(uv - vu), \qquad (30)$$

where $\hbar$ is Dirac's notation for $h/2\pi$.[14] It is easily shown that the new definition of $[u, v]$ satisfies all the previous operational rules of the Poisson brackets (6)–(9), plus the Jacobi identity (25).

Hamilton's canonical equations could now be written in the form

$$q_i = [q_i, H] = \frac{1}{i\hbar}(q_i H - H q_i),$$

$$p_i = [p_i, H] = \frac{1}{i\hbar}(p_i H - H p_i), \qquad (31)$$

and indeed, more generally

$$F(q_i, p_i) = \frac{1}{i\hbar}(FH - HF). \qquad (32)$$

In classical mechanics it is tautological but consistent to add the bracket equations (22) which express the condition of a canonical transformation, since the motion *is* in fact a succession of canonical transformations. In Dirac's theory the side conditions

$$q_i q_k - q_k q_i = 0, \quad p_i p_k - p_k p_i = 0, \quad q_i p_k - p_k q_i = i\hbar \delta_{ik}, \qquad (33)$$

are absolutely essential, because they express the commutation rules for the quantum mechanical variables q_i, p_i.

This re-interpretation of the classical equations of Hamiltonian dynamics opened a new vista in Dirac's thinking. He saw in quantum theory the emergence of a new type of algebra, which was *non-commutative*. He distinguished between two types of quantities: the '*q*-numbers' – which are non-commutative with respect to multiplication – and the '*c*-numbers', which are commutative; (ref. 15, see also ref. 17). Dirac realized that for a comparison between theory and experiment it becomes necessary to represent the *q*-numbers in terms of *c*-numbers (for example with the help of matrices or differential operators (cf. [16]). However, it may be possible to solve quantum mechanical problems purely in terms of a non-commutative algebra, without added tools. [This algebra possesses an added 'asterisk operation', defined by

$$(xy)^* = y^* x^*, \tag{34}$$

(corresponding to the 'Hermitian adjoint operator'), which restricts the q_i, p_i by the condition of 'self-adjointness'

$$q_i = q_i^*, \quad p_i = p_i^*. \tag{35}]$$

Indeed, Dirac succeeded in deriving the energy states of the hydrogen atom with the help of his algebraic algorithm.[15]

The 'canonical transformations', which played such an important role in Jacobi's work, could now be re-formulated in particularly simple form. The transformation

$$q_i' = S q_i S^{-1}, \quad p_i' = S p_i S^{-1}, \tag{36}$$

left the basic equations unchanged and thus corresponded to Jacobi's canonical transformations. [The condition (35) restricted the q-number S by the condition

$$S^{-1} = S^*. \tag{37}]$$

Today we have a clearer insight into the mutual relation of the ingenious ideas which erected the edifice of quantum mechanics from apparently so different viewpoints. What happened in all these efforts was that the 'phase space' of classical mechanics was replaced by the 'Hilbert space' of quadratically integrable functions. The Schrödinger equation gave a direct representation of the mechanical system with the help of the ψ-function which changed its position in Hilbert space according to a definite law. The eigenstates of the Schrödinger equation established a basic orthogonal reference system in Hilbert space, in which the ψ-vector and its motion in time could be analyzed.

The Born–Heisenberg–Jordan representation in terms of matrices gave another, but entirely equivalent description of the quantum mechanical system in terms of matrices, which could be defined by certain scalar products in Hilbert space, and calculable in terms of the orthogonal base vectors of that space, but obtainable without knowing the ψ-function. Instead of solving Schrödinger's wave equation, we can solve the matrix equations which are established by re-interpreting the Hamiltonian equations in terms of infinite matrices.

Finally Dirac's method of operating with a non-commutative algebra with given commutation rules can be conceived as operating again in the infinite-dimensional Hilbert space, but now operating *directly* with the vectors and tensors of that space, without setting up an orthogonal base system, which would resolve these vectors and tensors into their components. (For a more detailed survey of the early phases of quantum mechanics cf. the well-known books of Whittaker,[18] Jammer,[19] and Dugas[20].)

Something, however, was still hidden in all these formal correspondences and it was perhaps exactly *the* thing of fundamental significance. Today group theory belongs to the standard equipment of every quantum physicist. This was not so in the early days of quantum mechanics, when the great new evolution took place. The group theoretical viewpoint came into its full right only in the latest phase of quantum theory, which deals with the theory of elementary particles. There is, however, one great exception, in the person of E. Wigner, at that time research associate at the University of Goettingen, exactly the place, where S. Lie (in collaboration with F. Klein) developed around the end of the last century his fundamental investigations into the mathematical nature of continuous groups. Wigner was thoroughly familiar with group theory and thus he saw at once that the revision of classical mechanics by Dirac, Born and Schrödinger is in closest relation to group theory and can be completely formulated in group theoretical terms. The transformation (36) is one of the most fundamental operations of group theory, the commutator (30) arises from restricting this operation to the infinitesimal case, and the classical definition (5) of the Poisson bracket has likewise its group theoretical counterpart; (cf. ref. 21, p. 147). The Jacobi identity (25) is a fundamental must for the existence of an associative group.

Under these circumstances it is not surprising to find that Wigner saw the group-theoretical implications of the new quantum theory from its very beginnings and that in his profound papers he contributed to the deeper understanding of Schrödinger's discovery by adding to it the group-theoretical viewpoint.[22, 23]

The author expresses his deep-felt appreciation to the editor of this celebrating volume for giving him a chance to bring into focus in simple and unsophisticated language the vital and far-reaching contributions of Paul Dirac to the subject of the Poisson bracket. He wishes also to express his thanks to his colleagues at the School of Theoretical Physics, Dublin Institute for Advanced Studies, in particular to Professor L. O'Raifeartaigh, for many instructive discussions.

178 C. LANCZOS

REFERENCES

1. S. D. Poisson, *Journal de l'Ecole Polytechnique* **8**, 266 (1809).
2. J.-L. Lagrange, *Mémoires*, Institut de France (1808), p. 1.
3. J.-L. Lagrange, *Mémoires*, Institut de France (1808), pp. 257, 363.
4. *The Mathematical Papers of Sir William Rowan Hamilton*, vol. II, *Dynamics*, ed. A. W. Conway and A. J. McConnell (Cambridge University Press: 1940).
5. C. G. J. Jacobi, *Vorlesungen über Dynamik, 1842–1843* (Reimer, Berlin: 1866).
6. E. Whittaker, *Analytical Dynamics*, 4th ed. (Cambridge University Press: 1937).
7. H. Goldstein, *Classical Mechanics* (Addison-Wesley, Cambridge, Mass.: 1950).
8. C. Lanczos, *The Variational Principles of Mechanics*, 4th ed. (University of Toronto Press: 1970).
9. J. L. Synge, *Classical Mechanics, Encyclopedia of Physics*, III/I (Springer: 1960).
10. L. de Broglie, *Phil. Mag.* **47**, 446 (1924).
11. E. Schrödinger, *Ann. Physik* **79**, 361 (1926).
12. W. Heisenberg, *Z. Physik* **33**, 879 (1925).
13. M. Born, W. Heisenberg, P. Jordan, *Z. Physik* **35**, 557 (1926).
14. P. A. M. Dirac, *Proc. Roy. Soc. (London)* **109**, 642 (1925).
15. P. A. M. Dirac, *Proc. Roy. Soc. (London)* **110**, 561 (1926).
16. P. A. M. Dirac, *Proc. Roy. Soc. (London)* **111**, 281 (1926).
17. P. A. M. Dirac, *Proc. Cambridge Phil. Soc.* **23**, 412 (1926).
18. E. Whittaker, *History of the Theories of Aether and Electricity*, vol. II, *The Modern Theories 1920–1926*. (Th. Nelson and Sons: 1953).
19. M. Jammer, *The Conceptual Development of Quantum Mechanics* (McGraw-Hill: 1966).
20. R. Dugas, *Histoire de la Mécanique* (Griffon, Neuchatel: 1950).
21. L. O'Raifeartaigh, 'Unitary representations of Lie groups in quantum mechanics', in *Lecture Notes in Physics* **6** (Springer: 1970).
22. E. Wigner, *Z. Physik* **40**, 492 (1926).
23. E. Wigner, *Z. Physik* **43**, 624 (1927).

II

La 'fonction' δ et les noyaux

L. Schwartz

C'est en 1935 que j'entendis parler pour la première fois de la fonction δ; j'étais étudiant, et un camarade venait d'entendre une conférence de physique théorique, et m'en a parlé en ces termes: 'Ces gens-là introduisent une soi-disant fonction δ, nulle partout sauf à l'origine, égale à $+\infty$ à l'origine, et telle que $\int\delta(x)dx = +1$. Avec des méthodes de ce genre, aucune collaboration n'est possible.' Nous y avons un peu réfléchi ensemble, et avons abandonné; je n'y ai plus repensé jusqu'en 1945. A ce moment, c'est dans un but tout-à-fait différent que j'ai défini les distributions. J'étais tourmenté par les 'solutions généralisées' d'équations aux dérivées partielles. Si nous considérons l'équation des cordes vibrantes,

$$\frac{\partial^2 u}{\partial x^2} - \frac{1}{v^2}\frac{\partial^2 u}{\partial t^2} = 0,$$

on écrit sa solution générale sous la forme $u(x, t) = f(x+vt) + g(x-vt)$, où f et g sont deux fois dérivables; si elles ne sont que continues ou une fois dérivables, on a bien cependant l'impression que la fonction u définie ci-dessus est quand même solution, dans un certain sens, de l'équation des cordes vibrantes. D'où l'idée de solution généralisée. J'en avais eu besoin pour un problème, résolu dans un cas particulier, par Deny et Choquet. La définition des solutions généralisées est simple: u est solution généralisée, si toutes ses 'régularisées', $u*\phi$ (où ϕ est une fonction suffisamment différentiable pour que $u*\phi$ ait les dérivées voulues, et à support compact), est solution de l'équation. Cette définition avait d'ailleurs déjà été introduite antérieurement par Bochner. Quelque chose restait insatisfaisant; on pouvait ainsi dire que u était solution généralisée de l'équation des cordes vibrantes, mais ni $\partial^2 u/\partial x^2$ ni $(1/v^2)(\partial^2 u/\partial t^2)$ n'avaient de sens séparément. C'est de là que sont sorties les distributions. C'est seulement après que je me suis aperçu qu'elles donnaient la solution des difficultés rencontrées dans la fonction de Dirac; celle-ci devenait la distribution de Dirac. J'ai alors

regardé un certain nombre de travaux de physique théorique, et me suis
aperçu avec effroi de l'énorme 'percée' qu'avaient faite les physiciens
dans la manipulation des distributions, sans que les mathématiciens
leur en 'donnent le droit'. La physique théorique était pleine de distributions fort complexes, notamment les 'fonctions singulières' de la
physique quantique, distributions invariantes par le groupe de Lorentz ;
tout un chapitre de la physique s'était développé avec le plus grand
succès, sans que les fondements en soient assurés. Certaines de ces
fonctions singulières donnèrent encore, par la suite, bien du mal aux
mathématiciens (et aux physiciens, qui mirent du temps à accepter
l'impossibilité de la multiplication des distributions). Toute cette ex-
périence a été très instructive, tant pour les physiciens que pour les
mathématiciens. Cependant les leçons n'en ont pas suffisamment été
tirées, et les contacts entre les deux sciences sont encore restés trop
rares.

Ce n'est pas seulement pour la fonction de Dirac elle-même que
Dirac s'était lancé en avant, ni même pour toutes les fonctions singulières ; il avait l'idée des distributions comme noyaux. Considérons une
fonction de deux variables, $(x, y) \rightarrow K(x, y)$; elle définit une transformation fonctionnelle, faisant correspondre à la fonction f une fonction
g par $g(x) = \int K(x, y) f(y) \, dy$. La théorie des équations intégrales avait
donné un grand développement à ces noyaux. Mais toute transformation
ne peut se représenter ainsi ; si l'on veut écrire l'identité, le noyau devrait
être précisément $\delta(x-y)$, à cause de $f(x) = \int \delta(x-y) f(y) dy$. Dirac
écrivit alors que, si l'on accepte les fonctions singulières comme noyaux,
toutes les transformations fonctionnelles usuelles, même les opérateurs
differentiels, peuvent se représenter par des noyaux. Il se trouve en
effet que, si $K_{x,y}$ est une distribution à deux variables, et si ϕ est
une fonction C^∞ à support compact, on peut lui faire correspondre
une distribution T, notée $K.\phi$, qu'on peut symboliquement écrire
$T_x = \int K_{x,y} \phi(y) \, dy$, et que K définit ainsi une application linéaire continue de $\mathscr{D}_y$ dans $\mathscr{D}'_x$; en outre le théorème des noyaux dit précisément
que toute application linéaire continue de $\mathscr{D}_y$ dans $\mathscr{D}'_x$ peut s'écrire,
d'une manière unique, à partir d'un noyau distribution. Ce théorème,
que j'ai démontré en 1950, est directement inspiré de la lecture de
Dirac.

Cette idée de noyau, où la distribution de Dirac joue un rôle fonda-
mental, comme noyau de l'identité, réintervient constamment partout.
Je l'ai retrouvée récemment dans l'étude des désintégrations de mesures.
Soit (Ω, λ) un espace probabilisé, et soit f une fonction ≥ 0 borélienne

sur Ω. Soit $\mathcal{T}$ une sous-tribu de la tribu λ-mesurable. On définit alors une espérance conditionnelle de f par rapport à $\mathcal{T}$, comme suit: c'est une fonction $f^{\mathcal{T}} \geq 0$, appartenant à la tribu $\mathcal{T}$, et telle que, pour tout ensemble $A \in \mathcal{T}$, les intégrales $\int f^{\mathcal{T}}(\omega)d\lambda(\omega)$ et $\int_A f(\omega)d\lambda(\omega)$ soient égales. Cette espérance conditionnelle existe, et est unique à un ensemble λ-négligeable près.

On aimerait en fait avoir mieux: les probabilités conditionnelles. Le système des probabilités conditionnelles s'appelle aussi la désintégration de la mesure λ par rapport à la sous-tribu; c'est une famille de probabilités sur Ω, $(\lambda_\omega^{\mathcal{T}})_{\omega \in \Omega}$, indexée par Ω lui-même (donc, pour tout $\omega \in \Omega$, $\lambda_\omega^{\mathcal{T}}$ est une probabilité sur Ω), telle que $\omega \to \lambda_\omega^{\mathcal{T}}$ appartienne à la tribu $\mathcal{T}$, et que, pour tout $A \in \mathcal{T}$, l'intégrale de mesures $\int_A \lambda_\omega^{\mathcal{T}}d\lambda(\omega)$ soit le produit χ_A de λ par la fonction caractéristique de A. L'existence et l'unicité de la désintégration ont été demontrées par Jirina. Si on possède une désintégration, elle donne des espérances conditionnelles pour toutes les fonctions f boréliennes ≥ 0, d'un seul coup: on peut prendre $f^{\mathcal{T}}(\omega) = \lambda_\omega^{\mathcal{T}}(f) = \int_\Omega f(\omega')d\lambda_\omega^{\mathcal{T}}(\omega')$. Elle joue le rôle de noyau. Poussons plus loin. Soit F une fonction sur Ω, borélienne ≥ 0, non plus à valeurs numériques, mais à valeurs mesures: $(Y, \mathcal{Y})$ est un ensemble Y muni d'une tribu $\mathcal{Y}$, et, pour tout ω, $F(\omega)$ est une mesure ≥ 0 sur $(Y, \mathcal{Y})$. On pourra alors chercher une espérance conditionnelle de F par rapport à la sous-tribu $\mathcal{T}$; c'est une nouvelle fonction à valeurs mesures sur $(Y, \mathcal{Y})$, $F^{\mathcal{T}}$, appartenant à la tribut $\mathcal{T}$, et telle que pour tout $A \in \mathcal{T}$, les intégrales de mesures $\int_A F^{\mathcal{T}}(\omega)d\lambda(\omega)$ et $\int_A F(\omega)d\lambda(\omega)$ soient égales. Moyennant des conditions de compacité sur $(Y, \mathcal{Y})$, on peut prouver l'existence de l'espérance conditionnelle, et son unicité à un ensemble λ-négligeable près. La liaison entre espérances conditionnelles et désintégrations devient alors bilatérale. La désintégration $\omega \to \lambda_\omega^{\mathcal{T}}$ n'est autre que l'espérance conditionnelle, relative à la sous-tribu $\mathcal{T}$, de la fonction à valeurs mesures $\omega \to \delta_{(\omega)}$ (où, pour $\omega \in \Omega$, $\delta_{(\omega)}$ est la mesure formée de la masse $+1$ au point ω; le $\omega \to \delta_{(\omega)}$ est encore, si Ω est $\mathbb{R}$, le fameux $\delta(x-y)$); son existence résulte donc de l'existence des espérances conditionnelles des fonctions à valeurs mesures. Et inversement, si F est une fonction à valeurs mesures, son espérance conditionnelle relative à $\mathcal{T}$ peut s'obtenir à partir de la désintégration, toujours par la formule $F^{\mathcal{T}}(\omega) = \int_\Omega F(\omega')d\lambda_\omega^{\mathcal{T}}(\omega')$.

Employons les notations des distributions ou de la physique. Au lieu de $d\lambda(\omega)$, écrivons $\lambda(\omega)d\omega$. Les formules deviennent les suivantes. La fonction à valeurs mesures $\omega \to \delta_{(\omega)}$ devient $\delta(\omega, \omega')\,d\omega'$, avec $\int_\Omega \delta(\omega, \omega')\phi(\omega')d\omega' = \phi(\omega)$. La fonction F à valeurs mesures sera

$F(\omega, y)dy$, son espérance conditionnelle $F^{\mathscr{T}}$ sera $F^{\mathscr{T}}(\omega, y)dy$, avec la relation intégrale

$$\int_A F^{\mathscr{T}}(\omega, y) \, \lambda(\omega)d\omega = \int_A F(\omega, y) \, \lambda(\omega)d\omega.$$

La désintégration $\omega \to \lambda_\omega{}^{\mathscr{T}}$ se notera $\lambda^{\mathscr{T}}(\omega, \omega')d\omega'$, ou $\delta^{\mathscr{T}}(\omega, \omega')d\omega'$, avec l'égalité

$$\int_A \lambda^{\mathscr{T}}(\omega, \omega')\lambda(\omega)d\omega = \int_A \delta(\omega, \omega')\lambda(\omega)d\omega = \chi_A(\omega')\lambda(\omega').$$

Et $F^{\mathscr{T}}$ sera donnée à partir de F et de la désintégration par

$$F^{\mathscr{T}}(\omega, y) = \int_\Omega F(\omega', y)\lambda^{\mathscr{T}}(\omega, \omega')d\omega'.$$

On trouve bien alors, pour $A \in \mathscr{T}$:

$$\int_A \delta^{\mathscr{T}}(\omega, y)\lambda(\omega)d\omega = \int_A \lambda(\omega)d\omega \int_\Omega F(\omega', y)\delta_N(\omega, \omega')d\omega'$$
$$= \int_\Omega d\omega' F(\omega', y) \int_A \delta^{\mathscr{T}}(\omega, \omega')\lambda(\omega)d\omega$$
$$= \int_\Omega F(\omega', y)\chi_A(\omega')\lambda(\omega')d\omega' = \int_A F(\omega', y)\lambda(\omega')d\omega.$$

On peut étudier ensuite les désintégrations régulières par rapport à une famille de tribus $(\mathscr{T}^t)_{t \in R}$, dépendant du temps t, et obtenir des résultats, avec la même liaison bilatérale, pouvant s'appliquer fructueusement aux martingales et aux processus de Markov. Les noyaux pour définir des transformations, un noyau de Dirac pour l'identité, c'est encore la même idée, dans un contexte pourtant très différent.

Je m'excuse d'apporter ici une contribution purement suggestive à l'œuvre de Dirac, qui par ailleurs dépasse considérablement la découverte de la fonction δ !

12

On the Dirac Magnetic Poles

Edoardo Amaldi and Nicola Cabibbo

1. INTRODUCTION

Dirac's paper on 'Quantized singularities in the electromagnetic field',[1] read today, forty years after its publication, is still a subject of admiration and surprise. It was received by the Royal Society on 29 March 1931, i.e. more than one year before the first announcement by C. D. Anderson of the discovery of positive electrons.[2]

The paper opens with a page of epistemological flavour in which Dirac notices that it was expected 'by the scientific workers of the last century' that the mathematics required for the formulation of physical theories:

would get more and more complicated, but would rest on a permanent basis of axioms and definitions, while actually the modern physical developments have required a mathematics that continually shifts its foundations and gets more abstract. Non-Euclidean geometry and non-commutative algebra, which were at one time considered to be purely fictions of the mind and pastimes for logical thinkers, have now been found to be very necessary for the description of general facts of the physical world. It seems likely that this process of increasing abstraction will continue in the future and that the advance in physics is to be associated with a continual modification and generalization of the axioms at the base of the mathematics rather than with a logical development of any one mathematical scheme on a fixed foundation.

At the beginning of the second page Dirac notices that his previous paper on 'Electrons and protons', published about one year before,[3] 'may possibly be regarded as a small step according to this general scheme of advance'. He then recalls that an unoccupied state in the sea of uniformly filled negative kinetic energy states 'would appear to us as a particle with positive energy and positive charge' which, as he had originally suggested, 'should be identified with the proton'. He proceeds, summarizing the results of Weyl[4] about the value of the mass of such a hole and of Tamm and others[5] about the instability of matter due to annihilation of holes with ordinary electrons. He concludes, following Oppenheimer's suggestion,[6] that such a hole should be 'a new kind of particle unknown to experimental physics' and adds: 'We may call such a particle an anti-electron.' After a few considerations on the possibility of observing the creation of electron-anti-electron pairs by gamma rays, and their annihilation, he points out that 'the protons on the above views are quite unconnected with electrons' but 'presumably they will have their own negative states . . . an unoccupied one appearing as an anti-proton'.

After these introductory considerations Dirac states, at the beginning of the third page, that 'the object of the present paper is to put forward a new idea which is in many aspects comparable with this one about negative energies. It will be concerned essentially, not with electrons and protons, but with the reason for the existence of a smallest electric charge.'

In an attempt to reach a deeper understanding of the electromagnetic interaction, Dirac proposed a generalization of the formalism of quantum mechanics consisting of allowing for wave functions with non integrable phase. In order to preserve the probabilistic interpretation, the path dependence of the phase difference between two points (in space or space-time) must be the same for all wave functions and is, therefore, related to the dynamics of the system. Dirac shows that the non integrability of the phase can be interpreted in terms of the presence of an electromagnetic field, which does emerge naturally in the new formalism.[7]

In the new formalism the possibility also emerges of singularities of the electromagnetic field corresponding to single magnetic poles. Their strength g, however, is restricted to a multiple of a minimum value g_0:

$$g = n g_0 \quad (n = 0, \pm 1, \pm 2, \ldots) \tag{1}$$

$$g_0 = \frac{1}{2} \frac{\hbar c}{e^2} e = \frac{137}{2} e. \tag{2}$$

These relations provide a basis for the explanation of electric charge quantization. Reading Dirac's paper one has the impression that he was not entirely satisfied since he had hoped to understand the charge quantization without introducing a new entity.

The discovery of the possible existence of Dirac monopoles re-establishes a formal symmetry between electric and magnetic forces.[8] The small value of e implies a very large value of the unit of magnetic charge g_0 so that the empirical asymmetry between electricity and magnetism receives a quantitative explanation. Dirac also comments on the difficulty of creating pole pairs due to their strong binding (section 4).

During the forty years that have passed since the publication of Dirac's first paper many experimentalists have looked for magnetic poles in cosmic rays as well as among the secondary particles produced by high energy accelerators. In the present article we try to summarize only the work done on this subject after 1967 since the papers published before have been reviewed some time ago by one of the present authors.[9]

As everybody knows, no experimental evidence has been found up to the present for the existence of Dirac magnetic poles; only upper limits for their production cross-section in high-energy collisions have been established, the values of which depend on the charge and mass adopted for these new objects, but are, generally speaking, always very low.

In section 2 we will outline the main theoretical developments of Dirac ideas and in section 3 the more recent techniques and experiments aimed at the detection of Dirac poles. Section 4 presents a recent idea which could provide an explanation of the fact that up till now free Dirac poles have not been observed, even if they do actually exist.

2. AN OUTLINE OF THE MAIN THEORETICAL DEVELOPMENTS

Many papers have been published on the theory of Dirac monopoles since 1931. They can be roughly divided into two main groups, the first of which investigates the consistency of a theory where both electrically and magnetically charged particles are present. The second group includes papers of phenomenological nature, which attempt to gain an understanding of the behaviour of Dirac monopoles, necessary for the design of experiments aiming at their detection.[10, 9] Some of the more recent papers of the second group will be discussed in sections 3.1 and 4.

In the present section we give a brief outline of the purely theoretical developments, which have been reviewed in more detail in two excellent papers by Wentzel[11] and Zumino.[12]

In his 1931 paper, Dirac had discussed the motion of a single electric particle in the field of a magnetic pole. He had also outlined the method of solution of the corresponding Schrödinger equation. This problem has been further studied by Tamm,[13] Fierz,[14] Banderet,[15] Harish-Chandra,[16] Zwanziger[17] and others.[18] The main interest of these papers, as has been emphasized by Zumino, lies in the fact that at this level the theory is found to be completely consistent.

A typical new phenomenon which appears is the existence of an angular momentum along the line joining the pole to the electric particle. Assuming the electron and the monopole to be spinless particles, the total angular momentum is found to be[19]

$$\mathbf{J} = \mathbf{r} \times \mathbf{p} + \frac{eg}{c}\frac{\mathbf{r}}{r}. \tag{3}$$

If g has its smallest value allowed by equation (1) (i.e. $n = 1$), the additional term corresponds to a spin $\frac{1}{2}$.[20]

The wider problem is to formulate a second quantization theory of electric and magnetic particles as a generalization of quantum electrodynamics. A serious difficulty is due to the fact that the main tool for the study of normal quantum electrodynamics, i.e. the perturbative expansions, is precluded not only by the large value of the coupling constant

$$\frac{g^2}{\hbar c} = \frac{137}{4} n^2, \tag{4}$$

but also by the following considerations. If a perturbative expansion in e and g were consistent, it would be so for any value of these quantities, while we know that the theory can only be consistent if the Dirac relationship (1–2) is satisfied.

A formulation of field theory of electric and magnetic particles was given by Cabibbo and Ferrari.[21] This is based on Mandelstam's[22] formulation of quantum electrodynamics, which is essentially a second quantization version of the formulation of single particle quantum mechanics with non integrable phase, used by Dirac in his 1931 paper (section 1).

Cabibbo and Ferrari have shown that, just as in the case of a single particle formulation given by Dirac, Mandelstam's formulation of path dependent quantum electrodynamics allows for the presence of magnetic monopoles, if equations (1–2) are fulfilled. The treatment of electric and magnetic particles is then completely symmetric.

A different approach was started by Dirac in his second paper of 1948.[23] Here Dirac gives a complete quantum mechanical treatment for a system containing an arbitrary but fixed number of poles and charges. He starts from an action principle from which he derives the equations of motion in Hamiltonian form, and these are quantized by standard methods.

To this aim Dirac has to introduce the electromagnetic potentials. This can be done in all space, except for a set of lines (strings), one issuing from each pole. The coordinates describing the position of the strings are dynamical variables, but condition (1–2) again ensures the consistency of the formulation by making these variables ignorable. This theory, however, is incomplete, since it does not allow for a number of phenomena such as the creation of pairs or their annihilation.

The problem of deriving a complete quantum field theory from an action principle, has been solved by Schwinger.[24] The Dirac condition (1–2) plays again a dominant role in ensuring the covariance of the theory.

In his important papers Schwinger argues that the minimum value of n appearing in equation (1) should be a multiple of 4.[25] This follows from two different arguments each contributing a factor 2. The first one arises from the fact that in defining the action integral, Schwinger uses two strings for each monopole, obtaining in this way a more symmetric expression. A simple derivation of the second factor 2 is the following. The Dirac condition (1–2) is equivalent to asking that the magnetic flux through any closed surface should be a multiple of

$$2\pi \frac{\hbar c}{e} . \qquad (5)$$

This is indeed just the form in which the Dirac condition arises in his 1931 paper as well as in the field theoretical version of ref. 21: a pole of strength g lying inside a given closed surface contributes a flux $4\pi g$, so that equation (5) leads directly to (1–2). If, however, one allows the pole to lie exactly on the surface, its contribution to the magnetic flux would be $2\pi g$ which, combined again with equation (5), would lead to (1–2) with the further restriction that n should be even.

Both arguments do not appear to be completely cogent.[26] Since, however, we are aware that the versions given above are necessarily very sketchy, the interested reader is referred to Schwinger's original papers.[24]

The possible existence of particles endowed with both an electric and a magnetic charge was first envisaged by Eliezer and Roy,[27] who solved the Schrödinger equation of an electron in the field of one such particle. This possibility has been discussed by various authors.[17, 28, 29, 30] Particles of this type can be used for constructing schemes in which they appear as components of hadrons, strong interaction being naturally interpreted as a consequence of the large value of g.[31, 32, 29, 30]

A particularly elegant scheme is the dyon model proposed by Schwinger. A dyon is essentially a quark endowed with a magnetic charge.

In order to discuss the properties of dyons one has to generalize Dirac condition (1–2). This is simply done by taking into account the symmetry of Maxwell equations

$$(a) \quad \text{rot } \mathbf{H} - \frac{1}{c}\dot{\mathbf{D}} = \frac{4\pi}{c}\mathbf{J}_e$$

$$(b) \quad \text{div } \mathbf{D} = 4\pi\rho_e$$

$$(c) \quad \text{rot } \mathbf{E} + \frac{1}{c}\dot{\mathbf{B}} = -\frac{4\pi}{c}\mathbf{J}_m \qquad (6)$$

$$(d) \quad \text{div } \mathbf{B} = 4\pi\rho_m$$

under the generalized duality transformation[33]

$$
\begin{aligned}
\mathbf{E} &= \mathbf{E}' \cos\phi + \mathbf{H}' \sin\phi & \mathbf{H} &= -\mathbf{E}' \sin\phi + \mathbf{H}' \cos\phi \\
\mathbf{D} &= \mathbf{D}' \cos\phi + \mathbf{B}' \sin\phi & \mathbf{B} &= -\mathbf{D}' \sin\phi + \mathbf{B}' \cos\phi \\
\rho_e &= \rho_e' \cos\phi + \rho_m' \sin\phi & \rho_m &= -\rho_e' \sin\phi + \rho_m' \cos\phi \\
\mathbf{J}_e &= \mathbf{J}_e' \cos\phi + \mathbf{J}_m' \sin\phi & \mathbf{J}_m &= -\mathbf{J}_e' \sin\phi + \mathbf{J}_m' \cos\phi .
\end{aligned}
\qquad (7)
$$

This transformation represents a rotation by an angle ϕ in a two-dimensional space. The charge of each particle is represented by a vector in this space with components q_e and q_m. The Dirac condition can be stated in invariant form

$$q_e^1 q_m^2 - q_e^2 q_m^1 = n\hbar c, \qquad (8)$$

where the upper indexes 1 and 2 refer to two different dyons. Following Schwinger (who, however, would write the right-hand side of equation (8) with an extra factor 4) let us assume that there is a set of dyons D_i with charges $(e_i, m_i g_0)$, where g_0 is the smallest possible magnetic charge given by equation (2). Let us also assume that one of them has its magnetic charge exactly equal to g_0. We can select this to be the first one: $m_1 = 1$. From equation (8) applied to the dyons D_1 and D_i we obtain

$$e_i g_0 - e_1 m_i g_0 = n_i \hbar c = n_i e g_0, \qquad (9)$$

hence $\qquad\qquad\qquad e_i = m_i e_1 + n_i e.$

The electric charge of a dyon is thus seen to be the sum of a term proportional to its magnetic charge and a multiple of the electron charge. In this way, composite particles which are magnetically neutral ($\Sigma m_i = 0$) have an electric charge which is a multiple of e. The electric charge e_1 is, however, undetermined. It can be determined if we further assume that it is possible to form composite particles which are electrically neutral. This requires that for a certain set of values of the index i (any value of i possibly appearing more than once in the set) one has

$$(\Sigma m_i)e_1 + (\Sigma n_i)e = 0 \qquad (10)$$

(Σm_i) cannot be zero, since we require the composite particle to have a non zero magnetic charge; therefore equation (10) implies that e_1 and e have rational ratio. Schwinger considers in particular the case when the values of e_i/e are $\frac{2}{3}$ and $-\frac{1}{3}$ in close parallelism with the quark model.

If on the other hand, particles with fractional charge e_q and zero magnetic charge should exist, the minimum allowed value of g would be correspondingly larger, since, according to equation (2), the product $e_q g$ should remain equal to $\frac{1}{2}\hbar c$.

3. NEW EXPERIMENTAL SEARCHES FOR DIRAC MONOPOLES

Dirac's proposal of the existence of magnetic monopoles has stimulated a great number of experiments, some of which are based on very ingenious schemes. Many of them require the knowledge of the behaviour of magnetic poles in their interaction with bulk matter and external electromagnetic fields.

All properties of monopoles related to the long-distance behaviour of their electromagnetic interaction, can be obtained, with a satisfactory level of confidence, by applying the Dirac duality to the corresponding properties of electrically charged particles and taking into account the differences in masses and coupling constants. The Dirac duality[1] is the particular case of transformation (7) for $\phi = \pi/2$

$$\mathbf{E} \to \mathbf{H} \qquad\qquad \mathbf{H} \to -\mathbf{E}$$
$$\rho_e \to \rho_m \qquad\qquad \rho_m \to -\rho_e$$
$$\mathbf{J}_e \to \mathbf{J}_m \qquad\qquad \mathbf{J}_m \to -\mathbf{J}_e$$
$$\epsilon \to \mu \qquad\qquad \mu \to \epsilon$$

where, as usual, $\mathbf{D} = \epsilon\mathbf{E}$ and $\mathbf{B} = \mu\mathbf{H}$.

Among these properties are the interaction of monopoles with static and slowly varying electromagnetic fields, ionization energy losses,[34, 35] Čerenkov radiation,[36] transition radiation and, to a lesser extent, bremsstrahlung.[35, 10]

On the contrary the behaviour of a monopole once stopped in bulk matter cannot be inferred with the same degree of reliability, since it involves unknown properties of poles such as their short-distance interaction with nuclei.[10, 37]

In particular those experiments which are based on the diffusion of slow magnetic poles through matter or their extraction from pieces of magnetic materials by means of magnetic fields would lose significance if poles can bind strongly to nuclei.

3.1. RECENT DEVELOPMENTS OF DETECTION TECHNIQUES

During the last few years two new techniques for the detection of electrically charged particles have been developed and became of considerable practical interest. The first has already been applied to the search for magnetic poles, while the theory of the second has been developed to the point of becoming very promising for the search for very high energy monopoles.

The first of these techniques has been developed by Fleischer, Price and Walker;[38] it consists in the use of dielectric solids in which heavily ionizing particles produce tracks which can be revealed by preferential chemical etch. The method has some similarity to that of nuclear emulsions, but presents a few typical features worthy of mention:

(a) They are threshold detectors, the response of which is a function of energy loss through ionization.

(b) The flux of charged particles at minimum ionization that a plastic can stand without being damaged, is very high ($\sim 10^{12}$ particles/cm^2).

(c) The exposure can be very long since there are no fading effects of the damage produced by particles above the threshold.

(d) By calibrating the plastic and dosing the etching, one can obtain fairly good estimates for the value of the charge.

A semi-empirical theory of the formation of etchable tracks in dielectrics has been built up[39] assuming that the damage is produced by the secondary electrons of energetic charged particles within a critical cylinder around their path, where molecular fragments are formed which are more soluble than the parent molecule. The critical radius of the cylinder is taken to be about 20 Å, as is appropriate to the passage of the etchant along the track and the diffusion of reaction products back to the surface. At the critical radius the dosage approximates doses producing bulk damage under gamma irradiation. The calculations are in agreement with experimental results for heavy ions and have been extended to magnetic monopoles, which can be detected, also for relativistic velocities and $n \geqslant 1$ or 2, depending on the used dielectric material.

The other new technique is based on the observation of the so-called transition radiation, i.e. the radiation emitted from a charge crossing the interface between two media. The calculation of this effect was first performed by Frank and Ginsburg[40] and later by Garibian[41] and others[42] and investigated experimentally, in the case of electrons, by groups working at Erevan[43] and Brookhaven.[44]

The extension to the case of monopoles has been made by Mergellian[45] and Dooher.[46] The reader is referred to the original papers, in particular to that by Dooher for detailed developments and results. Here we may simply quote the result (similar to those holding for other phenomena: ionization, Čerenkov radiation, etc.) that, in the limit $\beta \rightarrow 1$, the transition radiation of monopoles is $(g/e)^2$ times larger than that of a charged particle. This means that the transition radiation produced when a magnetic monopole crosses the inter-media surface (e.g. metallic foils), can provide a useful signature for detecting such particles if they exist in cosmic rays or are produced either by accelerators or from cosmic ray interactions with nucleons.

In particular the well-known linearity in $\gamma(=E/mc^2)$ of the transition radiation intensity in the X-ray region, allows a relatively accurate mass measurement of a monopole if an independent energy measurement is made. Fig. 1 shows, for a few values of γ, the energy emitted by a magnetic pole, integrated over the solid angle in the forward direction, in the energy interval $d\omega$ for the X-ray region above 1 keV.

3.2. EXPERIMENTS WITH ACCELERATORS

A group at the Kurchatov Institute of Atomic Energy[47] has carried out an experiment at the Serpukov accelerator similar in principle to some of the experiments made a few years before at the Bevatron,[48] the CERN PS[49, 50] and the AGS.[51] The main advantage of the new experiment derives from the higher energy of the accelerated protons: a 70 GeV proton can produce pairs of Dirac poles of mass m_g up to 5 proton masses on a hydrogen target, and up to $7m_p$ on a heavier nuclei target. The poles produced in an aluminium target, placed inside the chamber, were accelerated transversally with respect to the circulating beam, by the magnetic field produced by one of the bending magnets of the accelerator, passed through the wall of the chamber and entered the detectors which were placed between the wall of the chamber and the pole pieces of the magnet. Each detector consisted of a tungsten plate, 3 mm thick, and 100 μm thick permandur foil, wrapped with polythene film. Eight 40×50 mm^2 detectors were put into the gap along the vacuum chamber, with the first one directly under the target. The tungsten plate was thick enough to stop completely monopoles of 35 GeV energy, which then diffuse in the permandur, where they remain trapped by the 'internal field'. The eight detectors covered a total 350 mm path, long enough for collecting and accumulating the monopoles emerging from the target.

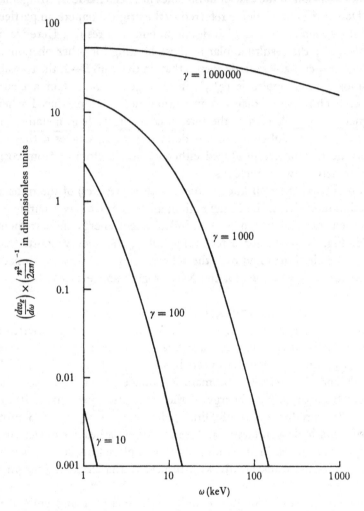

Fig. 1. Transition radiation spectrum from aluminium vacuum interface.

After irradiation the monopoles were extracted from permandur by a pulsed 220 kG magnetic field, and accelerated into nuclear emulsions. The magnetic field necessary for extracting from a ferromagnetic material a monopole of strength g is given, according to Goto et al.[52, 53] by the expression

$$H \geqslant H_0 = \pi M_0 \ln \left(\frac{g}{4\pi M_0 Z_0} \right), \tag{11}$$

where M_0 is the saturation magnetization ($4\pi M_0 = 24$ kG for permandur) and Z_0 a cut-off distance.

From equation (11) one obtains for permandur: $H_0 = 54$ kG for $n = 1$ and $H_0 = 68$ kG for $n = 12$. Since no track was observed which could have been attributed to monopoles, these authors placed an upper limit for the proton–nucleon cross-section of 1.5×10^{-41} cm²,[47] which later was pushed down to[54]

$$\sigma(95 \, \%) < 10^{-42} \text{ cm}^2 \text{ for } m_g \leqslant 7m_p. \tag{12}$$

Although the authors do not state for which range of values of n this result should hold, one can estimate, assuming equation (11) to be correct, that it should be valid from $n = 1$ up to rather large values of n. Gurevitch et al.[47] point out that the evaluation (12) is obtained assuming that the 10 kG field in the accelerator gap into which the target and the ferromagnetic traps were placed, did not lower in any way the efficiency of monopole accumulation.

It is difficult to evaluate the significance of this experiment because of the scanty information given by the authors about their estimates concerning the behaviour of monopoles in the various materials. This depends not only on the values of n and m_g, but also on the composition of the various materials through which the slow monopoles are expected to diffuse. In any case the experiment is based on the extraction of monopoles from a ferromagnetic collector, and therefore it involves the hypothesis that monopoles do not bind too strongly to nuclei present in the various materials.[10, 9, 37]

Other experiments should be made at the Serpukov machine using the external beam, so that the ferromagnetic trap will be out of the magnetic field, or, much better, by detecting them in flight, so that their slowing down and diffusion in matter would be avoided. A proposal of this type actually has been made by Blagov, Petukhov et al.,[55] who plan to collect the monopoles (produced in a target by an extracted proton beam) and bring them, by means of a magnetic channel, to a spark chamber spectrometer. The method has the interesting feature of allowing the detection of monopoles over a wide range of values of their magnetic strength, extending from rather large values of n to a fraction of the Dirac elementary monopole (2).

o

Table 1 summarizes all the experiments made up to now with accelerators. One more experiment is now in preparation by a group of CERN, Rome, Saclay and Strassburg[56] at the Intersecting Storage Rings (ISR) of the Meyrin Laboratory, in which the poles will be detected in flight by plastic track detectors. The advantage of such an

Table 1. *Accelerator searches of monopole*

Study by	Energy (Gev)	Max $\frac{m_g}{m_p}$	n	Upper limit of production cross-section quoted by authors
Bradner and Isbell[48]	6.3	1.1	1	2×10^{-35} cm^2
Fidecaro *et al.*[50]	27.5	3.0	1 to > 12	10^{-39} cm^2
Amaldi *et al.*[49]	25–28	3.0	1 and 2	6×10^{-41} cm^2 (95 %)
Purcell *et al.*[51]	30	3.0	1	2.1×10^{-40} cm^2 (95 %)†
Gurevitch *et al.*[54]	70	7.0	1 to > 12	10^{-42} cm^2 (95 %)

† The 95 % confidence limit given here is obtained by multiplying the 86 % confidence limit given by Purcell *et al.* (1.4×10^{-40} cm^2) by the factor 2.996/1.967. These numbers are the same as those appearing in equation (13) of section 3.3.

experiment is the fact that the energy in the c.m. is ~ 50 GeV so that pairs of poles of mass up to about 25 GeV can be produced, in principle, in proton–proton collisions. With the present luminosity of the ISR, if no pole is observed in a one year exposure, the upper limit that can be established on the proton–proton cross-section will be of the order of $10^{-37} \div 10^{-38}$ cm^2.

3.3. SEARCHES IN NATURE

An extensive search for monopoles in nature, has been made by Fleischer, Price, Schwarz, Woods and a few more collaborators, by using the solid-state track detectors developed by this same group (section 3.1).

In some researches, plastics, and in others, natural detectors were used, i.e. materials present on the Earth's surface in which heavily ionizing particles, such as fission products, heavy primaries of cosmic rays (and magnetic poles), produce damage along their track similar to those produced in plastics.

The various situations explored by these authors are shown in Fig. 2. The upper part (*a*) illustrates the type of phenomena underlying experiments made by various authors since Malkus' pioneering work.[57] A high-energy cosmic ray primary colliding with a nucleon in the upper atmosphere produces a pair of poles which are slowed down and pulled along the field lines of the Earth. They can be accelerated by

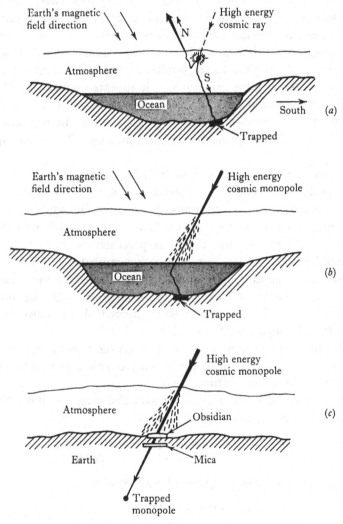

Fig. 2. Possible mechanisms that can lead to the observation of magnetic monopoles in nature: (*a*) they are produced by incoming high energy protons; (*b*) cosmic monopoles could be brought to rest either in the atmosphere or the ocean; (*c*) very high cosmic monopoles could penetrate deeply in the Earth.

a convenient magnetic field into a detector, or trapped in some ferro-magnetic or paramagnetic material, from which they are later extracted and detected.

The intermediate and lower parts of the same figure refer to the suggestion first put forward by Porter[58] who assumes that a large per-centage (possibly all) cosmic ray primaries with energy larger than 10^{17} eV could be Dirac poles. In Fig. 2b a very high-energy primary monopole is slowed down in the atmosphere and part of the ocean and then diffuses through the water to the ocean floor where it is trapped by ferromagnetic materials. In Fig. 2c the primary monopole which penetrates deeply into the Earth, produces a track in natural detectors.

Before discussing these experiments, it appears useful to recall that Porter's suggestion was advanced as a possible solution to the problem raised by the observation of extensive air showers in the range of $10^{18} \div 10^{20}$ eV, due to primaries incident isotropically on the upper atmosphere.

The existence of primaries of such high energies gives rise to two serious problems:[59] one is the mechanism of their acceleration, the other the problem of their containment inside the Galaxy.[60, 61]

The subject has undergone a remarkable evolution following recent discoveries: the pulsars and their interpretation as rotating neutron stars could provide an efficient acceleration mechanism.[62] The 2.7°K universal background radiation has, on the contrary, the effect of reducing the energy of fast particles, mainly through the inverse Compton effect.[61, 63] The problems mentioned above, however, are still far from being definitely settled.

Under these circumstances speculations on the possible existence of Dirac poles as a component of the primary cosmic rays should not be rejected without serious consideration.

Porter's proposal starts from the remark that magnetic poles placed in a magnetic field, gain energy at a rate of

$$n \times 300\tfrac{137}{2}H = n \times 2.055 \times 10^4 H \text{ eV/cm.}$$
$$(n = 1, 2, \ldots, H \text{ in G.})$$

If one takes for the average galactic field the value

$$H_G \simeq 3 \times 10^{-6} \text{ G,}$$

this would produce, over an interval of one light year, an energy gain

$$\Delta E = n \times 5.65 \times 10^{16} \text{ eV/l.y.}$$

Energies of the order of 10^{20} eV could be attained by a Dirac pole if the acceleration process continued over a distance of the order of one tenth of the galactic dimensions.

This over-simplified picture has been examined in more detail by Goto,[64] Alvarez[65] and more recently, by Osborn.[66] The last author established very low upper limits on the flux of monopoles by considering the experimental values of the frequency of muon-poor extensive showers. Osborn notices that a monopole of the very high energy suggested by the previous authors would produce by inverse Compton scattering on the 2.7 °K universal background radiation, a gamma ray spectrum in the range from 10^{13} to 10^{20} eV. These photons would produce muon-poor extensive showers. Therefore, using the experimental upper limits on the flux of photons in this energy range, derived from the study of muon-poor extensive showers, one can deduce an upper limit on the flux of monopoles. The values derived by Osborn are lower than those obtained from direct observation. Such an interesting argument does not, however, detract from the interest of direct experiments.

A figure of merit of any experiment of the type made by Fleischer et al. is provided by the product AT of the area A over which the poles are collected and the time of collection T: if the number of observed particles is zero, the corresponding $x = \xi$ % confidence limit for their flux is given by[9]

$$\Phi(x) = \frac{\ln(1-x)^{-1}}{AT} \ \mathrm{cm}^2\mathrm{s}^{-1}. \tag{13}$$

For example, a $x = 90$ % confidence limit is obtained from equation (13) by taking $\ln(1-x)^{-1} = \ln 10 = 2.303$.

In a first set of experiments, concerning the situation illustrated in Fig. 2a and b, two different types of potential collectors of monopoles deposited through the ages on the ocean floor were used: manganese nodules formed on the floor of the Southern Ocean[67] and ferromanganese deposits collected from the ocean floor at a position west of the Mid-Atlantic ridge (at 45° N latitude).[68] Both these deposits remained exposed on the ocean floor for long and well defined (16 ± 0.8 million years) periods of time, so that the product AT has a high value.

The experiments consisted of placing the samples in a magnetic field which, in the case of the manganese nodules, was pulsed and reached a maximum of about 265 kG, while, in the case of the ferromanganese deposits, was constant and equal to about 100 kG. In both cases the shape of the field lines was studied so that, for a very wide range of values of their mass, the extracted monopoles were focused on and pulled through plastic track detectors consisting of sheets of Lexan polycarbonate and Daicel cellulose nitrate, which differ in their detection threshold.

Here, again, one should add some word of caution about the significance of experiments of this type. They involve a number of assumptions about the behaviour of slow poles when they diffuse through different materials and are extracted from ferromagnetics.

A second set of experiments concerns the situation illustrated in Fig. 2c.[69, 70] The authors used a group of micas and obsidian samples, the age of which was already known or was measured from the stored tracks from spontaneous fission of uranium 238. Because neither mica nor obsidian would record tracks of monopoles with $n = 1$, the results of these experiments refer to monopoles with $n \geqslant 2$.

In none of these experiments were tracks observed which could be attributed to magnetic poles. Therefore only upper limits can be obtained for the flux of monopoles incident to the Earth or for their concentration in solid matter.

Table 2. *Collecting power of cosmic ray experiments*

	Authors	$A \times T$ (cm$^2 \times$ s)	Charge detectable (n)
Magnetic collection from atmosphere	Malkus[57]	10^{10}	< 1 to ~3
Magnetic collection from atmosphere	Carithers et al.[71]	6.9×10^{13}	< 1 to ~3
Extraction from magnetic outcrop	Goto et al.[53]	3×10^{12}	< 1 to ~3
Extraction from deep ocean sediment	Kolm et al.[72]	~4×10^{16}	⅓ to ~220
Extraction from Mn nodules	Fleischer et al.[67]	2.8×10^{14}	< 1 to ~120
Extraction from Mn crust	Fleischer et al.[68]	4.9×10^{17}	1 to ~60
Stored tracks in minerals	Fleischer et al.[69, 70]	2.3×10^{18}	2 to ∞

Table 2 summarizes the terrestrial searches for magnetic monopoles made up to now. The range of values of n is computed by Fleischer *et al.* by introducing into equation (11) the maximum value of the magnetic field used for extracting the monopoles. They also give for each experiment a range of values for the mass m_g of the poles, which has been estimated by a method not clearly explained in their papers. Fig. 3, taken from ref. 70, shows the ranges of magnetic charges and monopole

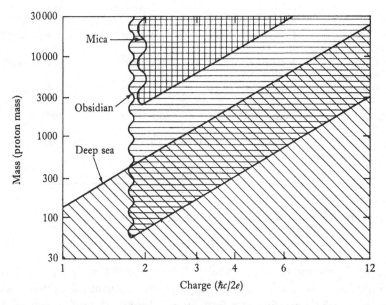

Fig. 3. Mass and magnetic charge domains that have been searched for monopoles by Fleischer et al.[70] According to these authors only the region of large masses and low magnetic charges ($n \leqslant 2$) are not covered by their experiments.

masses covered by the various experiments of Fleischer et al. Finally, Table 3 shows the 90 % confidence upper limits on the concentration of monopoles in solid matter.

In this table the results of two more experiments which aim to establish the concentration of magnetic poles in matter without extracting them from the sample also appear. This method has the advantage of not relying on current semi-qualitative theoretical considerations about the binding of monopoles to matter.

As pointed out by Alvarez[73] the generalized Faraday–Neumann law is immediately obtained from equation (6c)

$$\text{e.m.f.} = \oint \mathbf{E} \cdot \mathbf{dl} = -i_m - \frac{d}{dt}\,\Phi(\mathbf{B}), \tag{14}$$

Table 3. *Limits of the concentration of monopoles in solid matter (according to Fleischer* et al.)

		Amount of material (g)	90 % confidence limit on monopole concentration	charge range detectable (n)
Magnetic outcrop Adirondacks	Goto et al.[53]	$\sim 5 \times 10^3$	8×10^{-28}	< 1 to ~ 3
Mn nodules Southern Atlantic	Fleischer et al.[67]	31.5	1.3×10^{-25}	< 1 to ~ 120
Mn crust North Atlantic	Fleischer et al.[68]	7.7×10^3	5.2×10^{-28}	< 1 to ~ 60
From meteorite	Alvarez et al.[73]	$\sim 10^2$	4×10^{-26}	< 1 to ∞
Commercial copper	Vant-Hull[74]	6.5	6×10^{-25}	10^{-3} to ∞
Moon rocks Mare tranquilitatis	Alvarez et al.[77]	8.3×10^3	4.8×10^{-28}	$\frac{1}{8}$ to ∞
Ocean sediments	Kolm[72]	1.5×10^6	3×10^{-31}	$\frac{1}{8}$ to ~ 220

which, in absence of a variable flux of **B**, reduces to

$$\text{e.m.f.} = \oint \mathbf{E} \cdot \mathbf{dl} = -i_m. \tag{15}$$

If a total amount of a 'potential collector' containing N_g poles of a definite sign is carried by an appropriate mechanical system, for example a wheel rotating at a frequency f (hertz), through a coil of N turns, the magnetic current is given

$$i_m = N_g g f.$$

The d.c. voltage that should be observed according to equation (15) for $N_g = 1$, $f = 100$ Hz and $N = 2 \times 10^6$ is given by

$$\Delta V = n \times 0.82 \times 10^{-6} \text{ eV}$$

which can be measured without difficulty. This method was applied by Alvarez et al. to test an iron meteorite and various terrestrial potential collectors.[9] More recently it has been applied with a few technical improvements to the study of the Lunar rocks.

Before reporting on the main features and the significance of the negative result of this last experiment, one should recall the experiment of Vant-Hull[74] who used a superconducting interferometer with a magnetic flux sensitivity of $\sim 10^{-17}$ Wb ($= 10^{-9}$ maxwell).

As pointed out by Tassie,[75] if a Dirac magnetic pole with $n = 1$ is passed through the aperture of a superconducting ring, the e.m.f.(15) would increase the flux trapped by the superconducting ring by

$$g = 2\phi_0,$$

where

$$\phi_0 = \frac{h}{2e} \simeq 2 \times 10^{-5} \text{Wb} = 2 \times 10^{-7} \text{ maxwell}$$

is the quantum of flux. In the case of a superconducting quantum interferometer, the superconducting ring is broken by two Josephson junctions. The response of this device is periodic in quantum phase drop around the ring ($\Delta\gamma$), in the sense that the tunnelling junctions will support a zero-voltage current of maximum amplitude $I = I_0 \cos \Delta\gamma$. Hence, phase changes can be observed by monitoring the amplitude of the current.

Similarly the passage of a Dirac pole (with $n = 1$) through a quantum interferometer would cause the quantum phase to change by 4πrad, corresponding to the leakage of the flux $2\phi_0$ through the weak links.

With this instrument, Vant-Hull determined the absence of net magnetic charge (to $\pm 0.05\phi_0$) in single iron whiskers with an internal flux of $\sim 10^7 \phi_0$. A few samples were also tested in order to establish an upper limit for the magnetic charge of elementary particles.[76] The samples passed through the interferometer were: (1) a charge 7×10^{21} e; (2) a copper sample containing about 3×10^{24} electrons, protons and neutrons (~ 10 g); (3) tungsten and gold samples containing about 4.8×10^{24} electrons and protons and 7.2×10^{24} neutrons (~ 20 g).

From the measurements the magnetic charge on the neutron q^n_m and the difference ($q^p_m - q^e_m$) of the magnetic charges of the proton and the electron[76] were determined to be less in magnitude than 2×10^{-41} Wb or 10^{-26} Dirac elementary poles (2), while the magnitude of the magnetic charge on the electron was determined to be less than 8×10^{-39} Wb or 4×10^{-24} Dirac poles for electron. From these data it follows that in the absence of Dirac poles, one has

$$|\text{div } \mathbf{B}| = |\rho_m| < 1.2 \times 10^{-11} \rho_s \text{ Wb/m}^3$$

where ρ_s is the mass density of neutral matter. This result represents the lowest experimental value for div $\mathbf{B}$ derived from direct measurements. An even lower value could be obtained from the results of the experiment discussed below, but as far as we know, these data have not yet been analysed from this point of view.

The experiments by Alvarez et al.[77] mentioned above concern the search for Dirac poles in 8.37 kg of lunar surface material returned by the Apollo 11 crew. The lunar surface is considered to be the most (or one of the most) likely 'potential collectors' for monopoles, whether they belonged to primary cosmic rays or were produced in the collision of high-energy cosmic rays with nucleons of the lunar surface. In either case the lunar material would slow the monopole down, and trap it. The reasons in favour of the lunar surface include its great age and the small depth to which the surface has been churned during that long period of time. These two factors give the lunar surface the longest known exposure to cosmic rays. Finally, the absence of both an atmosphere and a magnetic field on the Moon allows a much better assessment of the fate of monopoles after they have been slowed down than in any material on Earth.

In this experiment the sample was transported around along a continuous path threading the winding of a coil, which was made of superconducting material and short-circulated by a superconducting switch. A small ˙current was stored in the superconducting circuit before the sample was run.

If a sample containing a monopole had been run, the induced e.m.f.(15) would have modified this current. After each sample had been circulated 400 times the superconducting switch was opened and a signal proportional to the current in the circuit was transferred electrically out of the cryostat, amplified and finally recorded on an oscilloscope.

The presence of a Dirac pole would have been detected as a difference between the signal obtained under these conditions and the one normally observed when the switch opened the 'standard current', that had been introduced as an overall check on this apparatus. The device was calibrated by means of a solenoid; a known change of its current gave rise to an e.m.f. in the superconducting coil through the second term on the right-hand side of equation (14). A statistical study of the signals obtained under these conditions showed that the measurement of the magnetic charge was affected by a 1 standard deviation error of about $\frac{1}{8}$ of the Dirac unit, when the sample was passed through the coil 400 times. Therefore a Dirac monopole with $n = 1$, was expected to produce a signal 8 times larger than 1 standard deviation.

The measured magnetic charge of 28 samples of a total mass of 8.37 kG was consistent with zero and statistically incompatible with the hypothesis that the absolute value of the magnetic charge was as large as or larger than, the elementary pole (1).

From these results Alvarez *et al.* set upper limits on the flux of monopoles present in the primary cosmic rays and on the number of monopoles produced by high-energy cosmic ray particles interacting with nucleons of the lunar surface material.

The actual values of both these upper limits depend upon unknown properties of the hypothetical magnetic monopole, such as its charge g, mass m_g and all the parameters that determine its range inside the lunar material before it comes to rest.

The authors express their results in terms of two parameters: the first one, indicated by n, relates the approximate range R, in g/cm², to the kinetic energy E, in GeV, by

$$R = 0.1(E/n^2).$$

For low velocities, when the monopole loses energy by ionization only, n is the magnetic charge measured in Dirac units. At higher energies, when the monopole starts to lose energy by bremsstrahlung, the effective value of n is expected to increase with E. The second parameter is the depth D, to which the lunar surface has been churned: two values of D have been considered, $D = 5$ cm corresponding to effectively no mixing depth, and $D = 100$ cm.

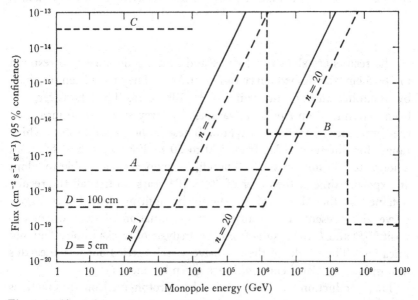

Fig. 4. 95 % confidence level upper limit on the flux of monopoles as a function of monopole energy obtained by Alvarez *et al.*[77] from measurements on lunar materials. The solid and dashed curves correspond to $D = 5$ and 100 cm, A is from ref. 67; B from ref. 69.

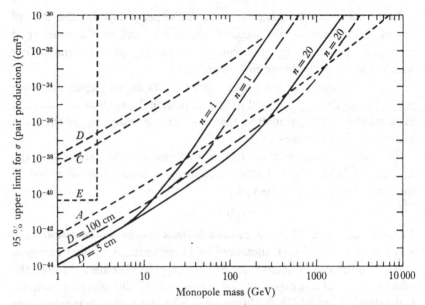

Fig. 5. 95 % confidence level upper limit on the monopole pair production cross-section in proton–nucleon collision obtained by Alvarez *et al.*[77] from measurements on lunar materials. The solid and dashed curves correspond to $D = 5$ and 100 cm. A is from ref. 68; C from ref. 69; D from ref. 79 and E from ref. 49.

The results are shown in Figs. 4 and 5: the solid curves correspond to $D = 5$ cm, the dashed curve to $D = 100$ cm. They have been obtained by assuming an exposure time of 3×10^9 years. This, however, has been determined on Apollo 11 samples by Urey *et al.*[78] who measured the abundance of spallation krypton. These authors found values which range, for different rocks, from 500 to 50 million years, and for fines amount to 330 million years. Therefore it appears reasonable to adopt an exposure time of the order of 300×10^6 years, so that all the results obtained by the Alvarez group should be multiplied by 10. In Fig. 4 curve A represents the results from examination of deep ocean deposit,[67, 68] and B the results from the analysis of tracks in obsidian and mica.[69, 70] The results of the extensive search carried out in the earth's atmosphere by Carithers *et al.*[71] are given by curve C.

The production of monopoles by proton–nucleon interactions depends upon the monopole pair-production cross-section. The curves of Fig. 5 are computed under the usual assumption that for each value of the monopole mass, σ_{pn} has a non-zero value which remains constant from threshold up to infinite energy of the incident proton. The flux of the primary cosmic ray of energy larger than E (in GeV) was assumed to be

$$\frac{1.4}{E^{1.67}} \text{ cm}^{-2}\text{s}^{-1}\text{sr}^{-1}.$$

Curves A and C represent limits for σ_{pn} obtained by Fleischer *et al.* from the study of deep ocean deposits[67, 68] and obsidian and mica.[69, 70] Curve D comes from a study of meteorites made by Petukhov and Yakimenko,[79] curve E corresponds to the accelerator study made by Amaldi *et al.*[49] The results of Gurevitch *et al.*[54] correspond to a straight horizontal line at 10^{-42} cm^2 extending on the right-hand side up to $m_g \simeq 7$ GeV.

Another upper limit concerning the difference of the numbers of monopoles of opposite sign present on the Moon has been derived by Schatten[80] from the analysis of measurements of the Moon magnetic field made by means of a magnetometer aboard Explorer 35. The result is 1.6×10^{-7} cm^{-3} or 7×10^{-32} per nucleon.

From all experimental results collected up to now about the upper limit on the flux of monopoles incident on the Earth or the Moon, one is brought to conclude that nature does not confirm Porter's assumption. This means that one cannot invoke Dirac monopoles for solving the problems arising from the observation of extensive air showers of energy larger than 10^{17} eV, but does not necessarily imply that monopoles do not exist.

Furthermore by considering the very low upper limits established for the production cross-section of pairs of monopoles shown in Table 1 and Fig. 5, one would be inclined to conclude that Dirac monopoles do not exist, or at least are extremely rare objects. Such a conclusion, however, would be too hasty. First of all the experiments and considerations presented here, or in previous papers, are based on pure guesses about many unknown properties of monopoles. Secondly, as we shall see in the next section, there is another possibility which would explain the extreme scarcity of monopoles in nature and the very low limits on production cross-section implied by all experiments carried out up to now.

4. POSSIBLE RE-ANNIHILATION OF POLE PAIRS CREATED IN HIGH-ENERGY COLLISIONS

The 1931 paper by Dirac ends with the remark that the very large attractive force between monopoles of opposite sign 'may perhaps account for why poles of opposite sign have never yet been separated'. In his second paper of 1948, Dirac points out that 'the forces of radiation damping must be very important for the motion of poles with an appreciable acceleration'.

The radiation damping is the basis of a recent proposal by Ruderman and Zwanziger,[81] which may explain why poles have not been observed until now. These authors suggest that when a pair of monopoles is produced, the radiation damping associated with their strong mutual attraction would lead to the loss of most of their kinetic energy, so that the pair would fall in and annihilate in a very short time. The observable effect of such a process would be the emission of a large number of photons.

Ruderman and Zwanziger argue as follows:

(a) At sufficiently large distances and low velocities the interaction between a pole–antipole pair is given by the Coulomb law

$$V(r) = g^2/r. \tag{16}$$

If we take $n = 2$, the Coulomb interaction is larger by a factor $\sim 2 \times 10^4$ than the electron–positron interaction. Vacuum polarization corrections can only alter the potential energy at any distance r, while the form factor of the poles, together with possible exchange of strongly interacting mesons, would modify the potential only for

$$r \lesssim r_0 = \frac{\hbar}{m_\pi c^2} \sim 10^{-13} \text{ cm}, \tag{17}$$

corresponding to the Compton wave length of the lightest hadron.

Moreover, in order for the vacuum to be stable against spontaneous production of pole pairs at distances down to $r \sim r_0$, the mass m_g of the poles should satisfy the condition

$$2m_g c^2 - \frac{g^2}{r_0} = 2m_g c^2 - \frac{g^2}{\hbar c} m_\pi c^2 \geqslant 0, \tag{18}$$

from which it follows

$$m_g \geqslant \frac{1}{2} \frac{g^2}{\hbar c} m_\pi = \frac{n^2}{4} 9.6 \text{ GeV}. \tag{19}$$

Because of this very large value of m_g the Compton wave length of a Dirac pole is smaller than that of the pion by the factor

$$\frac{r_0}{r_g} = \frac{n^2}{4} \frac{137}{2},$$

and the classical picture provides a rough but adequate description of the relative motion of the pole–antipole pair at distances $r \gtrsim r_0$, while it would completely fail if applied to the electron–positron pair for distances of the order of the classical radius of the electron ($\sim 2r_0$).

(b) In order that the pole–antipole pair can be separated, once they have been produced at a distance $\sim r_0$, their relative kinetic energy should be greater than

$$\frac{g^2}{r_0} \simeq 20 \text{ GeV}, \qquad (20)$$

even if all subsequent radiation is ignored. This means that the kinetic energy of their relative motion should satisfy the inequality

$$\frac{1}{2}\left[\frac{m_g c^2}{\sqrt{(1-\beta^2)}} - m_g c^2\right] \gtrsim \frac{g^2}{r_0},$$

which can be written in the form

$$\gamma-1 \gtrsim \frac{g^2}{r_0}\frac{2}{m_g c^2}. \qquad (21)$$

Therefore, unless m_g exceeds considerably the lower limit (19), the pole and antipole can be separated only if their relative velocity at $r=r_0$ is relativistic. But the coupling constant (4) of Dirac monopoles to photons is very large. Therefore one can expect (see point (c)) that in the creation process such a large fraction of the energy available would be radiated, that the residual kinetic energy of the two poles may not be sufficient for their separation.

(c) Nobody knows how to calculate the spectrum of the radiated photons or simply the energy they carry away.

The following argument shows, however, that a large energy loss should take place by emission of soft photons: and this can be estimated by classical considerations, at least for soft photons belonging to certain spectral regions.

The first assertion is justified by considering the probability that a pole pair is suddenly emitted with velocity $v \simeq c$, without an energy loss exceeding ΔE to any dipole photon. This is given approximatively by the expression

$$\exp\left[-\frac{8}{3\pi}\frac{g^2}{\hbar c}\ln\frac{m_g c^2}{\Delta E}\right] \simeq \exp\left[-\frac{2}{3\pi}n^2\,137\ln\frac{m_g c^2}{\Delta E}\right], \qquad (22)$$

obtained by replacing e with g in a standard expression used for computing radiative corrections. Equation (22) should hold when the energy $\hbar\omega$ given to any one photon is less than

$$\hbar\omega \simeq \frac{\hbar}{\tau} = \frac{\hbar c}{r_0} = m_\pi c^2 \qquad (23)$$

where $\tau = r_0/c$ is the acceleration time. Taking for m_g the limit value permissible according to equation (19) and for ΔE the value (23), the probability (22) turns out to be negligibly small.

For soft photons of wave length not large with respect to r_0 one can distinguish two emission regions: $r \lesssim r_0$ and $r > r_0$, in both of which the energy radiated by the pair of poles can be roughly estimated by classical considerations. These estimates confirm the idea that in order to separate a pole-pair, the energy available in their centre of mass should exceed $2m_g c^2$ by a very large factor.

The maximum distance to which the pole–antipole will be carried is expected to be of the order of 10^{-12} cm and the duration before returning to $r \lesssim r_0$ should be of the order of 10^{-22}–10^{-21} s. Finally the number of photons emitted is expected to be very roughly of the order of the coupling constant, i.e.

$$n \sim \geqslant n^2 \frac{g_0^2}{\hbar c} = \frac{n^2}{4} \, 137 \, .$$

This general picture provides a very reasonable explanation of the fact that up till now no experimental evidence has been found for the existence of isolated Dirac poles, even if they actually exist. Furthermore, as pointed out by Ruderman and Zwanziger, this mechanism would explain a few anomalous cosmic ray showers observed in stacks of nuclear emulsions exposed to cosmic rays at high altitude.[82, 83, 84] They consist of a number of electron pairs without any heavily ionizing particles. Fig. 6 shows the design of one of these events[83] consisting of 23 pairs and a 'trident' produced by one of the electrons of one of the pairs.

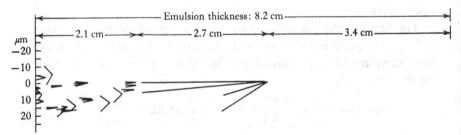

Fig. 6. Points of origin and opening angles of 23 pairs of the second event observed by De Benedetti *et al.*[83] in a stack of nuclear emulsions exposed at high altitude. The energy of the electron pairs range from about 50 MeV to 20 GeV and the distances of their points of origin, measured from that of the first pair, vary from 550 μm to about 4.7 cm. The radiation length in nuclear emulsion is 2.7 cm.

According to the authors[82, 83, 84] the probability that any of these events may be due to a fluctuation of a normal electron–photon cascade is extremely small, as one can easily recognize considering that the radiation length in nuclear emulsion is about 2.7 cm. These events can, however, be easily interpreted in terms of bursts of gamma rays of the kind envisaged by Ruderman and Zwanziger.

Newmeyer and Trefil[85] have applied the statistical model[86] to estimate the probability of producing a pair of poles in a high energy collision. By taking into account the interaction in the final state, they find that the emission of radiation reduces the probability of emission of free poles by two or more orders of magnitude, depending on the values of n and m_g.

These considerations indicate the opportunity for a search of typical multigamma events, which could be an indirect proof of the existence of magnetic poles. An experiment aiming to observe events of this type is in preparation by a Brookhaven–Rome group at the ISR of CERN.[87]

It is clearly very difficult to obtain reliable estimates of the cross-section for production of bound pairs of monopoles. A few very rough estimates, based on the idea of relating the cross-section for production of these objects to that for creation of pairs of muons, give, however, values which are large enough to conclude that the planned experiment could at least provide a significant upper limit for the cross-section.[87]

Thus, forty years after the publication of the 1931 Dirac paper the search for magnetic poles is pursued by many experimentalists inspired, or challenged, by the statement made by Dirac after having recognized that the existence of magnetic particles is a definite possibility within the formalism of quantum mechanics:[1] 'under these circumstances one would be surprised if Nature had not made use of it'.

REFERENCES
1. P. A. M. Dirac, *Proc. Roy. Soc. (London)* A133, 60 (1931).
2. C. D. Anderson, *Science* 76, 238 (1932).
3. P. A. M. Dirac, *Proc. Roy. Soc. (London)* A126, 360 (1930).
4. H. Weyl, *Gruppentheorie und Quantenmechanik*, 2nd ed. p. 234 (1931).
5. I. Tamm, *Z. Physik* 62, 545 (1930); J. R. Oppenheimer, *Phys. Rev.* 35, 939 (1930); P. A. M. Dirac, *Proc. Cambridge Phil. Soc.* 26, 361 (1930).
6. J. R. Oppenheimer, *Phys. Rev.* 35, 562 (1930).
7. P. A. M. Dirac acknowledges that the new formalism is not entirely new 'being essentially just Weyl's Principle of Gauge Invariance in its modern form'.
8. See equations (6) and (7) of section 2 and equation (11) in section 3.
9. E. Amaldi, 'On Dirac Magnetic Poles', in *Old and New Problems in Elementary Particles*, ed. G. Puppi, p. 1 (New York: 1968).

210   E. AMALDI AND N. CABIBBO

10. E. Amaldi, G. Baroni, H. Bradner, H. G. De Carvalho, L. Hoffmann, A. Manfredini and G. Vanderhaeghe, Report CERN 63–13, Nuclear Physics Division, 10 April 1963.
11. G. Wentzel, *Suppl. Progr. Theoret. Phys.* no. 37–8, 163 (1966).
12. B. Zumino 'Recent developments in the theory of magnetic charged particles' in *Strong and Weak Interactions*, ed. N. Zichichi, p. 711 (New York: 1966).
13. I. Tamm, *Z. Physik* 71, 141 (1931).
14. M. Fierz, *Helv. Phys. Acta* 17, 27 (1944).
15. P. P. Banderet, *Helv. Phys. Acta* 19, 503 (1946).
16. Harish-Chandra, *Phys. Rev.* 74, 883 (1948).
17. D. Zwanziger, *Phys. Rev.* 176, 1480 (1968).
18. H. J. Lipkin, W. I. Weissberger, M. Peshkin, *Ann. Phys.* (*N.Y.*) 53, 215 (1969).
19. In the classical book by J. J. Thomson, *Elements of the Mathematical Theory of Electricity and Magnetism* (Cambridge, Mass: 1st ed. 1900) it is shown (p. 396) that the superposition of the static fields generated by a charge $+e$ placed at a distance r from a magnetic pole $+g$, gives rise to an angular momentum eg/c (or $eg/4\pi$ in rationalized units with $c=1$) which is independent from the distance r and has the direction pointing from $+g$ to $+e$.
20. This point was stressed by M. N. Saha (*Indian J. Phys.* 10, 141 (1936)) who showed that the Dirac relationship expressed by (1–2) can be obtained by equating the angular momentum eg/c to an integral multiple of $\frac{1}{2}\hbar$. Later the same remark was made independently by other authors (H. A. Wilson, *Phys. Rev.* 75, 309 (1949); J. A. Eldridge, *Phys. Rev.* 75, 1614 (1949)). The role of this spin in monopole problems was pointed out by A. S. Goldhaber (*Phys. Rev.* 140B, 1407 (1965).
21. N. Cabibbo, E. Ferrari, *Nuovo Cimento* 23, 1147 (1962).
22. S. Mandelstam, *Ann. Phys.* (*N.Y.*) 19, 1 (1962).
23. P. A. M. Dirac, *Phys. Rev.* 74, 817 (1948).
24. J. Schwinger, *Phys. Rev.* 144, 1087 (1966); 151, 1022 (1966).
25. A minimum value of $n=2$ was suggested also by Alvarez who started from Saha considerations (ref. 20): if Dirac poles have spin $\frac{1}{2}$, the electron-pole system should be composed of two fermions and therefore n should be even, so that its minimum value should be 2. See ref. 65.
26. This view is shared also by Wentzel and Zumino (ref. 11 and 12).
27. C. J. Eliezer and S. Roy, *Proc. Cambridge Phil. Soc.* 58, ii, 401 (1962).
28. D. Zwanziger, *Phys. Rev.* 176, 1489 (1968).
29. J. Schwinger, *Phys. Rev.* 173, 1536 (1968); *Science* 165, 757 (1969); see in particular p. 227 of the book: J. Schwinger, 'Particles, Sources and Fields' (Addison–Wesley Publ. Co., Reading, Mass: 1970).
30. A. O. Barut, Review of hadron symmetries: mutliplets, supermultiplets and infinite multiplets. Preprint ETF-20-102E Kiev; Gables Conference on Fundamental Interactions at High Energy, p. 199, Gordon and Beach, 1970, 'Atoms and magnetic charges on models of hadrons' in *Topics in Modern Physics*, E. U. Condon volume (University of Colorado Press: 1971).
31. The precursor of these schemes is probably Saha (ref. 20) who suggested that the neutron could consist of bound pole-pairs.
32. R. A. Carrigan, *Nuovo Cimento* 38, 638 (1965).
33. Probably the first one to point out this was G. Y. Rainich, *Am. Math. Soc. Trans.* 27, 106 (1925), especially p. 120.

34. H. J. D. Cole, *Proc. Cambridge Phil. Soc.* **47**, 196 (1951).
35. E. Bauer, *Proc. Cambridge Phil. Soc.* **47**, 777 (1951).
36. J. M. Frank, *Vasilov Memorial Symposium* (USSR Acad. Sci., Moscow: 1952). D. R. Tompkins, *Phys. Rev.* **138B**, 248 (1965); **140B**, 433 (1965).
37. D. Sivers, *Phys. Rev.* D2, 2048 (1970).
38. P. B. Price and R. M. Walker, *J. Appl. Phys.* **33**, 3407 (1962); R. L. Fleischer and P. B. Price, *ibid.* **34**, 2903 (1963); *Science* **140**, 1221 (1963); R. L. Fleischer, P. B. Price and R. M. Walker: *Ann. Rev. Nucl. Sci.* **15**, 1 (1965); *Science* **149**, 383 (1965).
39. R. Katz and E. J. Kobetich, *Phys. Rev.* **170**, 401 (1967); M. Monnin: *International Topical Conference on Nuclear Track Registration in Insulating Solids and Applications* (Clermont-Ferrand, France: 1969) vol. ii, p. 73.
40. I. Frank and V. Ginsburg: *J. Phys. of USSR*, **9**, 353 (1945).
41. G. M. Garibian, *Zh. Eksperim. i Teor. Fiz.* **33**, 1403 (1957). *Soviet Phys. JETP* **6**, 33, 1079 (1958); *Zh. Eksperim. i Teor. Fiz.* **38**, 1814 (1960), *Soviet Phys. JETP* **11**, 1306 (1960). Erevan Institute of Physics Report TF-4(70): also obtainable in English translation by A. Maciulaitis as Gruman Research Department Translation TR-58.
42. M. L. Ter-Mikaelyan, *Dokl. Akad. Nauk SSSR* **134**, 318 (1960) (*Soviet Phys. Doklady* **5**, 1015 (1961)); M. L. Ter-Mikaelyan and A. D. Garazyan, *Zh. Eksperim. i Teor. Fiz.* **39**, 1693 (1960) (*Soviet Phys. JETP* **12**, 1183 (1969)); M. L. Ter-Mykaelyan, *Nucl. Phys.* **24**, 43 (1961).
43. A. I. Alikhanian, G. M. Garibian, K. A. Ispirian, E. M. Laziev and A. G. Oganession, in *Proceedings of the International Conference on High Energy Instrumentation* (Versailles, France: 1968).
44. L. C. L. Yuan, C. L. Wang, H. Uto and S. Prünster, *Phys. Letters* **31B**, 603 (1970); *Phys. Rev. Letters*, **23**, 496 (1969); **25**, 1513 (1970).
45. O. S. Mergellian, *Pubbl. Acad. Sci. Armenian SSR* **36**, (1963). Also obtainable in English translation by A. Maciulaitis as Gruman Research Department Translation TR-57.
46. J. Dooher, Gruman Research Department, Report RE-410J, June 1971.
47. I. I. Gurevich, S. Kh. Khakimov, V. B. Marthemianov, A. P. Mishakova, V. V. Ogwitzov, V. G. Tarasenkov, L. M. Barkov and N. M. Tarakanov, *Phys. Letters* **31B**, 394 (1970).
48. H. Bradner and W. M. Isbell, *Phys. Rev.* **114**, 603 (1959).
49. E. Amaldi, G. Baroni, H. Bradner, J. De Carvahlo, L. Hoffmann, A. Manfredini and G. Vanderhaeghe, *Proceedings of the Aix-en-Provence Conference* (1961), p. 155; *Nuovo Cimento* **28**, 773 (1963).
50. M. Fidecaro, G. Finocchiaro and G. Giacomelli, *Nuovo Cimento* **22**, 657 (1961).
51. E. M. Purcell, G. B. Collins, T. Fujii, J. Hornbostel and F. Tukot, *Phys. Rev.* **129**, 2326 (1963).
52. E. Goto, *J. Phys. Soc. Japan* **10**, 1413 (1959).
53. E. Goto, H. H. Kolm and K. W. Ford, *Phys. Rev.* **132**, 387 (1963).
54. I. I. Gurevitch, S. Kh. Khakimov, V. P. Marthemiamov, A. P. Mishakova, V. V. Ogwitzov, V. G. Tarasenkov, L. M. Barkov and N. M. Tarakanov, *Proceedings of the International Conference on Elementary Particles*, Amsterdam July 1–7, 1971.
55. M. I. Blagov, A. I. Isakov, V. A. Murashova, V. A. Petukhov, A. V. Samojlov, T. I. Syrejshchikova, Yu. Ya. Tel'nov, A. M. Frolov, Yu. D. Usachev and M. N. Yakimenko, Report no. 46 (1970) of the P. N. Lebedev Institute of Physics, translated in English at CERN by B. Hodge, CERN Translation 71-10 (March 1971).

56. CERN/ISRC/70-21, October 1970.

57. W. V. R. Malkus, *Phys. Rev.* **83**, 899 (1951).

58. N. A. Porter, *Nuovo Cimento* **16**, 958 (1960).

59. P. Morrison, 'The origin of cosmic rays', in *Handbuck der Physik*, ed. S. Flugge 45/1 (Berlin: 1961).

60. V. L. Ginsburg: *Soviet Astron. A. J.* **9**, 877 (1966).

61. S. I. Syrovatskii, *Proceedings of the 11th International Conference on Cosmic Rays*, p. 233 (Budapest: 1969).

62. E. Pacini, *Nature* **219**, 15 (1968); **224**, 160 (1969); J. E. Gunn and J. P. Ostriker, *Phys. Rev. Letters* **22**, 729 (1969); *Nature* **221**, 454 (1969); P. Goldreich and W. Julian, *Astrophys. J.* **157**, 869 (1969); J. P. Ostriker and J. E. Gunn, *Astrophys. J.* **157**, 1395 (1969); **160**, 979 (1970).

63. Kostantinov *et al.* Paper OG-101 presented at the Budapest Conference (ref. 61).

64. E. Goto, *Progr. Theoret. Phys.* **30**, 700 (1963).

65. L. W. Alvarez, Physics Notes from the Lawrence Radiation Laboratory, University of California, Memo 479 dated 11 July 1963.

66. W. Z. Osborn, *Phys. Rev. Letters*, **24**, 1441 (1970); **25**, 324 (1970).

67. R. L. Fleischer, I. S. Jacob, W. M. Schwarz and P. B. Price, *Phys. Rev.* **177**, 2029 (1969).

68. R. L. Fleischer, H. R. Hart jr., I. S. Jacob, and P. B. Price, W. M. Schwarz and F. Aumento, *Phys. Rev.* **184**, 1393 (1969).

69. R. L. Fleischer, P. B. Price and R. T. Woods, *Phys. Rev.* **184**, 1398 (1969).

70. R. L. Fleischer, H. R. Hart jr., I. S. Jacob, P. B. Price, W. M. Schwarz and R. T. Woods, *J. Appl. Phys.* **41**, 958 (1970).

71. W. C. Carithers, R. Stefanski and R. K. Adair, *Phys. Rev.* **149**, 1070 (1966).

72. H. H. Kolm, F. Villa and A. Odian, see ref. 9; H. H. Kolm *Phys. Today* **20**, 69 (1967).

73. L. W. Alvarez, Physics Notes from Lawrence Radiation Laboratory, University of California Memo 407, dated 9 March 1963.

74. L. L. Vant-Hull, *Phys. Rev.* **173**, 1412 (1968).

75. L. J. Tassie, *Nuovo Cimento* **38**, 1935 (1965).

76. This point was discussed by H. Harrison, N. A. Krall, O. C. Elridge, F. Fehsenfeld, F. Fite and W. B. Teutsch: *Am. J. Phys.* **31**, 249 (1963), who found as upper limit for the magnetic charge of protons and neutrons q_m^P, $q_m^n \lesssim 10^{-33}$ e.m.u. $\simeq 2 \times 10^{-24}$ e. See ref. 9.

77. L. W. Alvarez, P. H. Eberhard, R. R. Ross and R. D. Watt, *Science* **167**, 701 (1970).

78. K. Marti, G. W. Lugmair and H. C. Urey, *Science* **167**, 549 (1970).

79. V. A. Petukhov and M. N. Yakimenko, *Nucl. Phys.* **49**, 87 (1963).

80. K. H. Schatten, *Phys. Rev.* **D1**, 2245 (1970).

81. M. A. Ruderman and D. Zwanziger, *Phys. Rev. Letters*, **22**, 146 (1969).

82. M. Schein, D. M. Haskin and R. G. Glasser, **95**, 855 (1954); **99**, 643 (1955).

83. A. De Benedetti, C. M. Garelli, L. Tallone, M. Vigone and G. Wataghin, *Nuovo Cimento* **12**, 954 (1954); A. De Benedetti, C. M. Garelli, L. Tallone and M. Vigone, *Nuovo Cimento* **2**, 220 (1955).

84. M. Koshiba and M. Kaplon, *Phys. Rev.* **100**, 327 (1955).

85. J. L. Newmeyer and J. S. Trefil, *Phys. Rev. Letters* **26**, 1509 (1971).

86. R. Hagedorn, *Nuovo Cimento*, Suppl. **6**, 311 (1968); R. Hagedorn and J. Ranft, *Nuovo Cimento*, Suppl. **6**, 169 (1968).

87. CERN/ISRC/70-19 Part I and Part II; CERN/ISRC/70-19 Addendum.

13

The Fundamental Constants and Their Time Variation†

Freeman J. Dyson

1. DIRAC'S PROPOSAL AND SOME OTHERS

The following eight quantities are generally regarded as defining fundamental properties of the physical universe.

$c = 3 \times 10^{10}$ cm sec^{-1}, velocity of light.

$\hbar = 1.05 \times 10^{-27}$ erg sec, Planck's constant in Dirac's notation.

$e = 4.8 \times 10^{-10}$ erg$^{\frac{1}{2}}$ cm$^{\frac{1}{2}}$, unit of electric charge.

$m = 1.6 \times 10^{-24}$ gram, mass of the proton.

$g = 1.4 \times 10^{-49}$ erg cm^3, Fermi's constant of the weak interactions.

$G = 6.7 \times 10^{-8}$ erg cm gram^{-2}, constant of gravitation.

$H = 1.6 \times 10^{-18}$ sec^{-1}, Hubble's constant, new value measured by Sandage (1971).

$\rho = 10^{-31}$ gram cm^{-3}, mean density of mass in the universe.

In addition to these, there are several constants pertaining to nuclear and strong interactions, but it is not yet clear which strong-interaction constants should be regarded as fundamental. The numerical values quoted for the eight listed constants are of varying degrees of reliability. H is uncertain by about a factor 2. The value of ρ is determined from the density of observed mass, and could be larger by as much as a factor 10^3 if various forms of invisible mass are also present.

From the eight listed quantities, five dimensionless ratios can be formed,

$$\alpha = (e^2/\hbar c) = 7.3 \times 10^{-3},$$
$$\beta = (gm^2c/\hbar^3) = 9 \times 10^{-6},$$
$$\gamma = (Gm^2/\hbar c) = 5 \times 10^{-39},$$
$$\delta = (H\hbar/mc^2) = 10^{-42},$$
$$\epsilon = (G\rho/H^2) = 2 \times 10^{-3}.$$

† This article is written in honor of P. A. M. Dirac's 70th birthday. It discusses some of the consequences of Dirac's work. It does not pretend to be a complete review of the literature concerning variation of constants.

Dirac (1937, 1938) first raised the question which is the subject of this article. Which of these five ratios, if any, remain constant as the universe evolves?

The conventional view of the behavior of the constants may be defined as follows. The five ratios are divided into 'laboratory constants' α, β, γ, and 'cosmological constants' δ, ϵ. The cosmological constants are assumed to vary with the time t that has elapsed since the beginning of the universe, while the laboratory constants are not. Specifically, it is supposed that the structure of the universe is described by an open cosmology in which the distance between any two widely-separated galaxies varies linearly with t. The behavior of the five ratios is then given by

Hypothesis A (Conventional View):

$$\alpha, \beta, \gamma \text{ constant}, \quad \delta \sim t^{-1}, \quad \epsilon \sim t^{-1}. \tag{1.1}$$

There are four arguments against the conventional view, each based on aesthetic considerations rather than on observational evidence.

1.1

Dirac (1937, 1938) introduced the *numerological principle*, which states that very large or very small numerical coefficients ought not to occur *a priori* in the basic laws of physics. He observed that the numerical values of α, β, γ, δ, ϵ can be brought into harmony with the numerological principle if we assume that the two very small ratios γ, δ vary with t while the rest are constant. Specifically, he assumed

Hypothesis B (Dirac):

$$\alpha, \beta, \epsilon \text{ constant}, \quad \gamma \sim t^{-1}, \quad \delta \sim t^{-1}. \tag{1.2}$$

This hypothesis has two main consequences. It requires that all gravitational forces (measured by γ) decrease with time by a fraction of the order of 10^{-10} per year. It also requires a zero-curvature cosmology so that ρ shall vary with time like t^{-1} instead of t^{-3}.

1.2

Sciama (1953) and Dicke (1961) introduced *Mach's principle* into the discussion. Mach's principle has as many interpretations as it has interpreters. Roughly speaking, it asserts that all dynamical phenomena must be caused by matter and not by intrinsic properties of empty space. Sciama and Dicke interpret this to mean that the inertial mass

μ of any object is equal to the mass $(\mu\psi/c^2)$ which it acquires from the gravitational potential ψ generated by the rest of the mass in the universe. It follows that

$$\psi = c^2. \tag{1.3}$$

If we assume a homogeneous universe with radius r and density ρ, and neglect numerical factors such as $(4\pi/3)$, we find

$$\psi = G\rho r^2$$
$$H = cr^{-1},$$

and hence

$$\psi = (G\rho c^2/H^2) = \epsilon c^2.$$

Mach's principle in the form (1.3) then reduces to

$$\epsilon = 1. \tag{1.4}$$

This means that there must be just enough mass in the universe to make the space-time either closed or asymptotically flat. The conclusion (1.4) supports Dirac's Hypothesis B and is inconsistent with Hypothesis A.

The observational evidence is not in conflict with (1.4). Although the value of ϵ deduced from visible mass is only 2×10^{-3}, there are many forms of invisible mass which could bring the true value up to unity. Examples of invisible mass with which the universe may be filled are, in increasing order of undetectability, (1) ionized intergalactic gas, (2) stars of low luminosity, (3) dead galaxies, (4) pebbles, (5) black holes, (6) neutrinos and (7) gravitational waves.

Brans and Dicke (1961) have proposed a detailed theory of gravitation and cosmology based on Mach's principle. They assume that $G = \phi^{-1}$, where ϕ is a scalar field which can vary both in space and time. The theory contains one arbitrary parameter ω which is assumed to be of the order of magnitude 10. Solution of the cosmological equations gives

$$G \sim t^{-r}, \quad H \sim t^{-1}, \quad \rho \sim t^{r-2},$$

with $r = (2 + \tfrac{3}{2}\omega)^{-1}$. The Brans–Dicke theory thus leads to:

Hypothesis C (Brans–Dicke):

$$\alpha, \beta, \epsilon \text{ constant}, \quad \gamma \sim t^{-r}, \quad \delta \sim t^{-1}. \tag{1.5}$$

This differs from Dirac's Hypothesis B in that r is a small number of the order of 0.05. According to (1.5), gravitational forces decrease with time by a fraction between 10^{-12} and 10^{-11} per year.

1.3

Gamow (1967) revived interest in the question of variation of the constants by suggesting that α might vary as t. He argued that the natural quantity to discuss was not the quantum-mechanical constants α and γ but rather the ratio

$$(\gamma/\alpha) = (Gm^2/e^2) = 7.10^{-37} \tag{1.6}$$

between the gravitational attraction and the electrostatic repulsion of two protons. According to Dirac's numerological principle, (γ/α) should vary as t^{-1}. Gamow proposed that instead of letting γ vary we keep γ fixed and let α vary instead.

Hypothesis D (Gamow):

$$\beta, \gamma, \epsilon \text{ constant}, \quad \alpha \sim t, \quad \delta \sim t^{-1}. \tag{1.7}$$

1.4

Teller (1948), Landau (1955), DeWitt (1964) and Isham, Salam and Strathdee (1971) have presented a variety of arguments leading to a relation between electromagnetism and gravity of the form

$$\alpha^{-1} \sim \log \gamma^{-1}. \tag{1.8}$$

Teller suggested this on purely numerological grounds, as a possible way of explaining why α is of the order of magnitude 10^{-2} rather than of the order of unity. Landau, DeWitt and Salam introduced in different ways the physical idea that the fundamental gravitational length

$$a = (G\hbar/c^3)^{\frac{1}{2}} = 1.6 \times 10^{-33} \text{ cm} \tag{1.9}$$

provides a natural cut-off to the logarithmic divergences that exist in quantum electrodynamics. The length a is the limit below which all notions of distance become meaningless because of quantum fluctuations in the space-time metric. If space-integrals are cut off at the distance a, the divergent quantities of quantum electrodynamics typically become finite expressions of the form

$$Z = f(\alpha \log (\hbar/mca)) = f(\tfrac{1}{2}\alpha \log \gamma^{-1}), \tag{1.10}$$

where $f(x)$ is some function which can in principle be calculated as a power-series in x. Now various arguments can be invoked to lead to the conclusion that a self-consistent electrodynamics is possible only for some special value of Z. For example, Z might be the fraction of the electron mass that is electromagnetic in origin, and then $Z = 1$ would be a possible consistency condition. Alternatively, Z might be the ratio between unrenormalized and renormalized electromagnetic fields, and then $Z=0$ expresses the 'bootstrap' condition that unrenormalized fields do not exist in nature. Such arguments are admittedly vague, but they make plausible the idea that an equation

$$\alpha \log \gamma^{-1} = k \qquad (1.11)$$

should hold, with a numerical value of k which is of the order of unity and in principle calculable. That is to say, they give support to Teller's proposal (1.8).

If we take (1.8) together with Dirac's form of the numerological principle, we obtain as a variant of Hypothesis B,

Hypothesis E (Teller):

$$\beta, \epsilon \text{ constant}, \quad \alpha \sim (\log t)^{-1}, \quad \gamma \sim t^{-1}, \quad \delta \sim t^{-1}. \qquad (1.12)$$

The rate at which α changes with time is measured by the quantity

$$R = (\dot{\alpha}/\alpha). \qquad (1.13)$$

Hypotheses A, B, C predict $R=0$. Hypothesis D gives

$$R \cong +5 \times 10^{-11} \text{ year}^{-1}, \qquad (1.14)$$

while Hypothesis E implies

$$R \cong -5 \times 10^{-13} \text{ year}^{-1}. \qquad (1.15)$$

The remaining sections of this article discuss the observational evidence required for a decision between these various hypotheses.

2. EVIDENCE CONCERNING VARIATION OF α

2.1 ASTRONOMICAL EVIDENCE

Spectra of very distant objects give direct evidence concerning the behavior of atomic energy-levels in the remote past. All atomic transition energies are, in the non-relativistic approximation, proportional to the Rydberg energy

$$Ry = \tfrac{1}{2}(\mu e^4/\hbar^2) = \tfrac{1}{2}\alpha^2 \mu c^2, \qquad (2.1)$$

where μ is the electron mass. So long as all optical wavelengths have the same dependence upon α, any variation in α would multiply every wavelength by the same factor. Such a uniform shift of wavelengths cannot be distinguished observationally from a Döppler shift produced by motion of the source. If the non-relativistic approximation to the energy-levels were exact, astronomical spectra could give no evidence for or against a variation of α with time.

Fortunately, some of the observed lines have a relativistic fine-structure with splittings theoretically proportional to α^4 rather than to α^2. Bahcall and Schmidt (1967) have collected examples of spectra of remote objects in which fine-structure can be accurately measured. The best evidence comes from five radio-galaxies whose optical spectra include emission lines of doubly-ionized oxygen. The observed red-shifts lie between 0.17 and 0.26, indicating that the light has been emitted between 10^9 and 5×10^9 years ago. Since the sources are galaxies rather than quasars, there is little doubt that the estimates of distance are correct in order of magnitude. The oxygen lines have a fine-structure which shows no signs of departure from a uniform red-shift. In each of the five spectra, the ratio of fine-structure to measured wavelength is equal to the same ratio for a spectrum measured in the laboratory. Since the ratio of fine-structure to wavelength is proportional to α^2, Bahcall and Schmidt conclude that α can have varied by at most a fraction 3×10^{-3} of itself during the last 2×10^9 years. Therefore

$$|R| \leq 2 \times 10^{-12} \text{ year}^{-1}, \tag{2.2}$$

and Gamow's Hypothesis D, which implies (1.14), is excluded. However, the astronomical evidence is not able to exclude Teller's Hypothesis E.

2.2 GEOPHYSICAL EVIDENCE

If we have a long-lived radioactive nucleus whose decay-rate λ depends on α, then the abundance of the nucleus in nature can tell us something about the behavior of α over the period since the nucleus came into existence. We define the sensitivity of the nucleus to be

$$s = \frac{\alpha}{\lambda}\frac{d\lambda}{d\alpha}, \tag{2.3}$$

so that λ varies with α^s when α changes slightly. Other things being equal, nuclei of highest sensitivity will provide the most precise evidence concerning variation of α. Geophysical observations give direct information about the quantity

$$L = (\dot{\lambda}/\lambda). \tag{2.4}$$

This is related to the rate of change of α defined by (1.13) according to

$$L = sR. \tag{2.5}$$

The three kinds of decay-activity which can be studied are alpha, spontaneous fission, and beta. These were analyzed by Wilkinson (1958), Gold (1968) and Peebles and Dicke (1962) respectively.

2.3 ALPHA-DECAY

The Gamow theory of alpha-decay gives a well-known formula for the decay-rate, depending on the penetration of the Coulomb barrier by the alpha-particle. A rough approximation to the Gamow formula is

$$\lambda = \Lambda \exp\left[-4\pi Ze^2/\hbar v\right], \tag{2.6}$$

for an allowed transition, where Z is the charge-number of the product nucleus, v is the alpha-particle velocity, and Λ is a function depending less sensitively on e and v. The decay energy is

$$\Delta = 2mv^2. \tag{2.7}$$

According to (2.3), the sensitivity is

$$s = -4\pi Z\alpha(c/v)(1 - \tfrac{1}{2}u) \tag{2.8}$$

with

$$u = \frac{\alpha}{\Delta}\frac{d\Delta}{d\alpha}. \tag{2.9}$$

To calculate u, we assume that the energy of any nucleus can be divided into two parts

$$E = E_N + E_c, \tag{2.10}$$

where E_N represents nuclear potential and kinetic energy and is independent of α, while E_c is the Coulomb energy

$$E_c = Z^2 A^{-\frac{1}{3}}\alpha \times 80 \text{ Mev}. \tag{2.11}$$

These assumptions are open to debate, but they are supported by a large amount of empirical data concerning nuclear masses. From (2.11) we deduce

$$u = [(Z+2)^2(A+4)^{-\frac{1}{3}} - Z^2 A^{-\frac{1}{3}}][0.6 \text{ Mev}/\Delta]. \tag{2.12}$$

The most favorable example of a long-lived alpha-active nucleus is uranium 238, for which $\Delta = 4.2$ Mev. In this case, (2.8) and (2.12) give

$$u = 8, \tag{2.13}$$

$$s = +500. \tag{2.14}$$

The positive variation of λ arising from the increase of v with α outweighs the negative variation of λ arising from the factor e^2 in the Gamow penetration formula.

We have to look at the geophysical evidence to see how much the decay-rate of U_{238} may have varied in the past. For this purpose we require old rocks in which we can reliably measure (a) the abundance of U_{238}, (b) the abundance of Pb_{206} produced by U_{238} decay, and (c) a value for the age, independent of the uranium–lead evidence. It is difficult to avoid circularity in an argument which uses age-estimates of rocks to test for a variation of constants which would destroy the validity of the age-estimates. However, a large body of evidence indicates that the U_{238} decay-rate has remained within 20 % of its present value over the last 2×10^9 years. We do not discuss the evidence here in detail, because it will turn out to be less crucial than the evidence concerning beta-decay. For Uranium 238 we have

$$|L| \leq 10^{-10} \text{ year}^{-1}, \tag{2.15}$$

and therefore (2.5) and (2.14) give

$$|R| \leq 2 \times 10^{-13} \text{ year}^{-1}. \tag{2.16}$$

This upper limit on R is ten times as precise as the limit obtained from the radio-galaxy spectra, and is formally inconsistent with (1.15). But, in view of the many uncertainties involved in the argument which led to (2.16), it cannot be said that the evidence from alpha-decay excludes Teller's Hypothesis E with certainty.

2.4 SPONTANEOUS FISSION
The theory of spontaneous fission is similar to that of alpha-decay, the decay-rate being again dominated by a Gamow penetration factor of the form (2.6). One might expect that the sensitivity s would be greater for spontaneous fission because the charge Z of the alpha-decay product is now replaced by the product of the charges of the two fission products. However, the effect of the greater effective charge is outweighed by the effect of the greater energy-release Δ in spontaneous fission. Gold (1968) finds for the spontaneous fission of U_{238}

$$s = +120. \tag{2.17}$$

The geophysical evidence concerning spontaneous fission in ancient rocks is remarkably precise, since the fission fragments leave characteristic tracks of ionization-damage in certain types of rock, and these tracks can be made visible by etching. The main uncertainty in the spontaneous fission rate comes from the uncertainty in the age of the rock. The observations indicate that for spontaneous fission of U_{238},

$$\lambda = 7 \times 10^{-17} \text{ year}^{-1}, \tag{2.18}$$

with a variation of at most 10 % over the last 2×10^9 years. We thus obtain

$$|L| \leq 5 \times 10^{-11} \text{ year}^{-1} \tag{2.19}$$

and

$$|R| \leq 5 \times 10^{-13} \text{ year}^{-1}, \tag{2.20}$$

confirming with somewhat less precision the result obtained from alpha-decay.

2.5 BETA-DECAY

The rate of beta-decay depends on the decay energy Δ according to a power-law

$$\lambda = \Lambda \Delta^p, \tag{2.21}$$

where the exponent p depends on details of the decay process. For low-Z nuclei with an l-times-forbidden decay, we have

$$p = l + 3 \tag{2.22}$$

for electron-emission and

$$p = 2l + 2 \tag{2.23}$$

for electron-capture decay. For high-Z nuclei with small Δ, the exponent becomes

$$p = 2 + (1 - \alpha^2 Z^2)^{\frac{1}{2}}, \tag{2.24}$$

independent of l. Neglecting the slight dependence of p on α, the sensitivity becomes

$$s = pu, \tag{2.25}$$

with u given by (2.9). For the details of the beta-decay theory see Konopinski (1943) and Berényi (1968).

We estimate u by means of (2.10) and (2.11) as before, and obtain

$$u = \pm (2Z + 1) A^{-\frac{1}{3}} [0.6 \text{ Mev}/\Delta], \tag{2.26}$$

where the plus sign refers to an electron-capture transition $(Z + 1 \to Z)$, and the minus sign refers to an electron-emission transition $(Z \to Z + 1)$. Table 1 is a list of the nuclei with long beta-decay half-lives which are candidates for detailed analysis. The table shows that rhenium 187 is uniquely sensitive as an indicator of variation of α, as Peebles and Dicke (1962) pointed out. It seemed at first that the Gamow penetration formula (2.6) would make U_{238} the most sensitive nucleus, but in fact the small value of Δ gives Re_{187} a thirty-fold advantage.

Since rhenium 187 is singled out as a diagnostic material by its purely nuclear properties, it is very lucky that this isotope also happens to have several fortuitous practical advantages as a material for geophysical observations.

(a) It is an abundant isotope, 63 % of natural rhenium.

(b) Its decay product osmium 187 is non-volatile (unlike argon 40) and will be retained in place after the decay.

(c) The natural abundance of osmium is not so large as to swamp the radiogenic Os_{187}.

(d) Rhenium occurs predominantly in iron-phase meteorites, which are in general superior to stony meteorites both in age and in chemical and physical integrity.

Table 1. *Long-lived beta-radioactive nuclei. The parameter* $s = (\alpha/\lambda)(d\lambda/d\alpha)$ *measures the sensitivity of the decay rate* λ *to variations of the fine-structure constant* α

Nucleus	Half-life (years)	Δ (MeV)	p	u	s
K_{40}	1.3×10^9	1.31	6	-5	-30
Rb_{87}	5×10^{10}	0.275	5	-36	-180
Te_{123}	1.2×10^{13}	0.06(EC)	6	210	1260
Re_{187}	4×10^{10}	0.0025	2.8	-6400	$-18\ 000$

2.6 EXPERIMENTS ON RHENIUM 187

There are three main sources of information about the decay of Re_{187}, (a) isotopic analysis of molybdenite ores, (b) isotopic analysis of iron meteorites, and (c) laboratory measurement of the decay-rate.

(a) Hirt, Tilton, Herr and Hoffmeister (1963) analyzed ten different molybdenite ores and measured the abundance-ratios

$$x = \frac{Os_{187}}{Os_{186}}, \quad y = \frac{Os_{187}}{Re_{187}}. \tag{2.27}$$

They found $x > 100$ in all these ores. The value of x in normal osmium is 0.8. The osmium in the ores is therefore at least 99 % radiogenic, a circumstance that makes the analysis of the measurements particularly simple. The age τ of each ore was measured by dating the adjacent rock-strata above and below it. Unfortunately there was no more direct way to date the ores independently of the evidence provided by the rhenium decay itself. The estimated ages of the ores ranged from 3×10^8 to 3×10^9 years. The errors in τ are of course less well controlled than the errors in y.

All ten ores gave results consistent with a linear relation

$$y = 1.6 \times 10^{-11} \, \tau, \qquad (2.28)$$

within the estimated error of the measurements. This implies a half-life

$$T = (4.3 \pm 0.5) \times 10^{10} \text{ years} \qquad (2.29)$$

for Re_{187}, averaged over the various times that have elapsed since the deposition of the ores.

(b) Herr, Hoffmeister, Hirt, Geiss and Houtermans (1961) measured the isotope ratios

$$x = \frac{Os_{187}}{Os_{186}}, \quad z = \frac{x}{y} = \frac{Re_{187}}{Os_{186}}, \qquad (2.30)$$

in fourteen iron meteorites. In this experiment the values of x are not large, indicating that the meteorites contain non-radiogenic osmium. The values of z range from 2 to 8, and over this interval the linear relation

$$x = 0.83 + 0.065z \qquad (2.31)$$

holds for all fourteen samples within the errors of measurement. The value $x = 0.8$ for normal osmium is also consistent with (2.31). The straight-line relation (2.31) is intelligible only if all the measured meteorites have the same age τ. Let T be the half-life of Re_{187} averaged over the age of the meteorites. The slope of (2.31) is then given by

$$b = 0.065 \pm 0.015 = 2^{(\tau/T)} - 1 \approx 0.7(\tau/T), \qquad (2.32)$$

and hence

$$T = (11 \pm 2)\tau. \qquad (2.33)$$

The main weakness of this determination of T is that we have no direct and independent measurement of τ. A number of lines of evidence converge upon a value of τ equal to the age of the earth

$$\tau = 4.5 \times 10^9 \text{ years.} \qquad (2.34)$$

If this is accepted, (2.33) gives

$$T = (5 \pm 1) \times 10^{10} \text{ years}, \qquad (2.35)$$

in agreement with (2.29). The meteoritic determination of T is less accurate than the ore determination but extends substantially further back into the past.

(c) Brodzinski and Conway (1965) measured the Re_{187} half-life by observing the decay directly in a proportional counter. They found the value

$$T = (6.5 \pm 0.5) \times 10^{10} \text{ years.} \qquad (2.36)$$

This result caused some excitement because it seemed inconsistent with (2.29). If one could take the difference between (2.29) and (2.36) to represent a true change in the decay-rate with time, it could be explained by a time-variation of α of magnitude

$$R \sim 3 \times 10^{-14} \text{ year}^{-1}. \qquad (2.37)$$

However, a variation as large as this would be inconsistent with the meteoritic half-life (2.35). If one believed that the α-variation given by (2.37) was real, he would also have to believe that the iron meteorites were much younger than the experts supposed.

Brodzinski and Conway proposed to explain their low value of the decay-rate in a different way. They quoted a theoretical calculation by Gilbert (1958) which indicated that 30 % of the Re_{187} decay-electrons should be emitted directly into bound states of the osmium atom. The decay mode

$$\text{Rhenium atom} \rightarrow \text{Osmium atom} + \text{neutrino} \qquad (2.38)$$

is invisible in a counter but contributes to the geophysically observed decay-rate. However, Bahcall (1961) recalculated the bound-electron process and obtained a result conflicting strongly with Gilbert. A reliable theory of the bound-electron decay is still lacking. This is an interesting theoretical problem on the borderline between physics and chemistry. At present one can only say that Bahcall's argument makes it highly unlikely that the fraction of Re_{187} decays in the bound-electron mode exceeds 1 %. Brodzinski and Conway's explanation of their long half-life is therefore unacceptable.

Like all low-energy beta-ray experiments, the Re_{187} half-life measurement is technically fiendish. The rhenium has to be in the counter in gaseous form. Volatile compounds of rhenium are few and tend to be chemically unstable. Brodzinski and Conway worked with biscyclopentadienylrhenium hydride $[(C_5H_5)_2ReH]$, which is barely volatile at 130 °C. They attempted to maintain this compound in the counter at its saturation vapor-pressure, but were frustrated by the fact that the stainless-steel wire catalysed its chemical decomposition. They measured an energy-spectrum of the decay electrons which had a quite different shape from that predicted theoretically. In these circumstances, it is not possible to assess reliably the accuracy of their half-life measurement.

Drever and Payne (1968) recently repeated the Brodzinski–Conway experiment with various technical improvements. The main changes were (*a*) elimination of exposed metal surfaces in the counter, and

(b) use of unsaturated instead of saturated vapor. They obtained an electron spectrum agreeing well with theory, and a value

$$T = (4.7 \pm 0.5) \times 10^{10} \text{ years} \tag{2.39}$$

for the half-life. The discrepancy with the geophysical half-life has completely disappeared.

From (2.29), (2.35) and (2.39) we deduce that the rhenium decay-rate did not change by more than 10 % in 10^9 years. Hence

$$|L| \leq 10^{-10} \text{ year}^{-1}, \tag{2.40}$$

and the final upper limit on the rate of change of α is

$$|R| \leq 5 \times 10^{-15} \text{ year}^{-1}. \tag{2.41}$$

This is the most precise of all results obtained so far concerning variation of natural constants. In particular, it decisively demolishes both Gamow's Hypothesis D and Teller's Hypothesis E.

3. EVIDENCE CONCERNING VARIATION OF THE WEAK-INTERACTION CONSTANT

In the analysis of the Re_{187} decay we have assumed that the Fermi constant β did not change with time. All beta-decay rates are proportional to β^2, and therefore the constancy of the rhenium decay-rate λ is consistent with varying α if β and α should happen to vary together in such a way that $\beta^2 \alpha^s$ is constant, with $s = -18\ 000$.

To separate the effects of variation of α and β, we must compare the decay-rates of two nuclei with widely different values of s, for example Re_{187} and K_{40} (see Table 1). The evidence for constancy of the K_{40} decay-rate is about as accurate as that for Re_{187}. The ratio of the K_{40} and Re_{187} decay-rates is thus (a) independent of β, (b) virtually as sensitive to α as the Re_{187} decay-rate alone, and (c) known to be constant within a few parts in 10^{10} per year. From the constancy of the ratio we can then conclude instead of (2.41)

$$|R| \leq 2 \times 10^{-14} \text{ year}^{-1}, \tag{3.1}$$

independent of any assumption concerning β. It follows from (3.1) and Table 1 that the effect of variation of α on the K_{40} decay-rate is completely negligible. The constancy of the K_{40} decay-rate then implies in turn

$$|\dot\beta/\beta| \leq 10^{-10} \text{ year}^{-1}. \tag{3.2}$$

Wilkinson (1958) has discussed the more direct evidence for constancy of β obtained from the study of ancient pleochroic haloes.

R

Pleochroic haloes (see Henderson (1934)) are spheres formed by alpha-ray tracks around specks of uranium-bearing mineral embedded in mica. The intensities of the haloes of different radii give information concerning the proportions of different daughter-activities in the decay chain from uranium to lead. Since some of the daughter-activities involve branching between alpha- and beta-decay, the haloes would show observable effects if the alpha–beta branching ratios in ancient times had differed from their present values. The haloes in fact show no evidence of any variation of the ratios. From this Wilkinson deduced an independent upper limit on $|\dot{\beta}/\beta|$, somewhat less precise than (3.2).

Some authors have speculated that Dirac's numerological principle might apply to β as well as to γ. Dirac's Hypothesis B would then be modified by assuming

$$\beta \sim \gamma^f \sim t^{-f} \tag{3.3}$$

where f is a small fractional exponent, for example $f = \frac{1}{4}$ or $f = \frac{1}{8}$. A variation of β of this magnitude is consistent with (3.2) and is not excluded by any known evidence.

If (3.3) were true back to very early times ($t \sim 10$ minutes) in the initial phase of a 'Big-Bang' cosmology, there would be a drastic effect on the primaeval helium abundance. Neutrons would have decayed rapidly to protons before being captured to form deuterium, and very little helium would have been formed until stars came into existence. However, the evidence for a substantial primaeval helium abundance is still equivocal, and the validity of the 'Big-Bang' model at such early times cannot be considered proved. So the helium abundance provides only suggestive but not decisive evidence against (3.3).

4. EVIDENCE CONCERNING VARIATION OF GRAVITATIONAL FORCES

The heart of Dirac's proposal (1.2) is the decrease of gravitational forces with time. Unfortunately the observational tests of (1.2) are much more difficult than the tests of variation of α, and our knowledge of the behavior of gravitational forces is by several orders of magnitude less precise than our knowledge of electromagnetic forces.

4.1 EFFECT OF VARYING γ ON PLANETARY ORBITS

Consider a planet of mass M_2 traveling around a sun of mass M_1 in a circular Newtonian orbit. If this system is treated quantum-mechanically, the radius of the orbit is

$$r = (n^2 \hbar^2 / G M_1 M_2^2), \tag{4.1}$$

and the total energy is

$$E = -(G^2 M_1^2 M_2^3 / 2n^2 \hbar^2), \qquad (4.2)$$

where n is the principal quantum number, just as in the theory of the hydrogen atom. When G varies slowly, the quantum number n does not change. According to the correspondence principle, n is also an adiabatic invariant in classical mechanics. It follows from (4.1) and (4.2) that orbital radii and energies vary as

$$r \sim \gamma^{-1}, \quad E \sim \gamma^2, \qquad (4.3)$$

if the gravitational coupling-constant γ is slowly changing. Dirac's proposal (1.2) therefore implies that the mean Earth–Moon and Earth–Sun distances increase linearly with t. More precisely, these distances will show a rate of secular increase

$$(\dot{r}/r) \sim 5 \times 10^{-11} \text{ year}^{-1}, \qquad (4.4)$$

in addition to other secular changes produced by tidal interactions. The Brans–Dicke Hypothesis C predicts a smaller rate of increase

$$(\dot{r}/r) \sim 3 \times 10^{-12} \text{ year}^{-1}. \qquad (4.5)$$

What possibilities exist for observing these secular motions? The Earth–Moon distance r is 4×10^{10} cm, which gives according to (4.4)

$$\dot{r} = 2 \text{ cm/year.} \qquad (4.6)$$

Laser reflectors have been emplaced on the Moon in order to measure r accurately. In addition to providing a wealth of information about the internal dynamics of the Earth–Moon system, the laser-ranging experiment can perhaps test the existence of a secular motion as small as (4.6). The present accuracy of the measurements of r is about 30 cm, but this is expected to improve by a factor of ten as the Earth-based end of the experiment is further refined. So far as experimental technique is concerned, an effect as small as (4.6) could probably be detected over the next five years. Unfortunately, the theoretical uncertainties in the Earth–Moon problem are more difficult to overcome. Tidal interactions produce secular changes of r much larger than (4.6), and these tidal effects depend critically on dissipative processes in the Earth's oceans and mantle which are hard either to observe or to calculate precisely. The Earth–Moon system historically resisted quantitative explanation for more than a century after Newton provided the basic theory. The Moon's motion will probably remain too much affected by theoretical ambiguities to be a decisive test for a prediction as delicate as (4.6).

In the long run, a better means of testing (4.4) is provided by inter-planetary ranging observations. The planets Venus, Mercury and Mars describe orbits which are undisturbed by tidal effects to the precision here required. The orbit radii are of order 2×10^{13} cm, so that (4.4) gives

$$\dot{r} \sim 10 \text{ meters/year.} \tag{4.7}$$

To test (4.7) we need ranging accuracy of the order of 10 m continued over several years. At present the errors in ranging to a radio transponder at interplanetary distances are of the order of 100 m, but an accuracy of 10 m could certainly be achieved in an experiment designed for the purpose.

Direct radar ranging to the planets will not give distances to the necessary accuracy, because of the unknown topography of their surfaces and the consequent diffuseness of the echoes. Shapiro and his colleagues (1971) have recently analyzed the results of a long series of radar observations of Mercury and Venus. Their tracking errors are of the order of 1 km, and their limit on the possible rate of variation of γ is

$$|\dot{\gamma}/\gamma| \leq 4 \times 10^{-10} \text{ year}^{-1}. \tag{4.8}$$

This is the most exact direct measurement of $\dot{\gamma}$ so far obtained.

An independent interplanetary probe carrying a radio transponder will also not be satisfactory as an indicator of changes in γ, because its orbit will be seriously perturbed by solar radiation pressure and by other non-gravitational forces. The necessary accuracy combined with predictability can only be achieved by a radio-transponder carried on a probe orbiting Mars or Venus. The measurements of distance to the probe then define the motion of the center of mass of the planet to the required precision, independently of minor perturbations in the orbit of the probe around the planet. After such observations have been continued for several years, a mammoth numerical integration of the equations of motion of the entire solar system must be performed, in order to disentangle the effects of varying γ from the effects of ordinary perturbation of planetary orbits by one another. Such a calculation was already done by Shapiro et al. (1971) in order to arrive at the result (4.8). If orbiting probes can be emplaced and the necessary calculations organized on a routine basis, a decision for or against the existence of the effect (4.7) should be obtainable within a few years. So Dirac's Hypothesis B is directly testable by the method of interplanetary ranging. On the other hand, to test the Brans–Dicke Hypothesis C would require accuracies of measurement and calculation substantially beyond anything that we yet know how to do.

4.2 EFFECT OF VARYING γ ON STELLAR EVOLUTION

Assume that a star is in a stable phase of its evolution, for example in the hydrogen-burning main-sequence phase, while γ is slowly varying. Assume also that the shape of the mass-distribution and temperature-distribution in the star does not vary, so that the evolution proceeds through a sequence of homologous configurations. Each configuration of the star is then determined by its radius r and central temperature T. The virial theorem implies that the variations of T, r, and G are related by

$$T \sim \gamma r^{-1}. \tag{4.9}$$

The density of any element of mass in the star varies as

$$\rho \sim r^{-3}, \tag{4.10}$$

and the opacity varies according to Kramers' law

$$\kappa \sim \rho T^{-7/2} \sim r^{-3} T^{-7/2}. \tag{4.11}$$

There are two ways to determine the luminosity L of the star. On the one hand, the law of thermonuclear energy generation gives

$$L \sim \rho T^n \sim r^{-3} T^n, \tag{4.12}$$

where $n \approx 4$ for the proton–proton reaction which is dominant in the sun, and $n \approx 17$ for the carbon-cycle reactions dominant in hotter main-sequence stars. On the other hand, the law of radiation transport gives

$$L \sim r^3 T^4 / \tau, \tag{4.13}$$

where τ is the mean diffusion time for a photon to travel from the center of the star to the surface. Since

$$\tau \sim (\kappa / rc), \tag{4.14}$$

(4.13) and (4.11) give

$$L \sim (r^4 T^4 / \kappa) \sim r^7 T^{15/2}. \tag{4.15}$$

Putting together (4.9), (4.12) and (4.15), we find

$$T \sim \gamma^{20\lambda}, \tag{4.16}$$

$$r \sim \gamma^{1-20\lambda}, \tag{4.17}$$

$$L \sim \gamma^{7+10\lambda}, \tag{4.18}$$

with

$$\lambda = (2n+5)^{-1}. \tag{4.19}$$

The remarkable feature of (4.18) is that L is highly sensitive to changes in γ but quite insensitive to the value of n. For the sun with $n = 4$, $\lambda = 0.07$,

$$L \sim \gamma^{7.7}, \qquad (4.20)$$

and so Dirac's Hypothesis B implies

$$L \sim t^{-7.7}. \qquad (4.21)$$

The flux F of solar radiation incident on the earth varies with Ld^{-2}, where d is the Earth–Sun distance. Hence (4.3) and (4.20) give

$$F \sim \gamma^{9.7}. \qquad (4.22)$$

The fact that F is so extremely sensitive to changes in γ makes it tempting to use the climatic history of the earth as evidence of the constancy of γ. For example, if Hypothesis B were true, F would have decreased by a factor of about 10 since the earth was formed. Such a large variation of F seems highly implausible, but it is not possible to say that it is excluded by our knowledge of climatic history, so long as we do not understand the basic determinants of climate. To make the argument conclusive, we need to know quantitatively the dependence upon F of the mean surface temperature of the earth. At present we cannot even say with certainty that an increase in F would not lead to a decrease in surface temperature, through some unforeseen redistribution of the earth's atmospheric circulation. Any simple statement of the type 'Life on earth could not have survived if F had had twice its present value a billion years ago' ignores the unpredictable richness of phenomena in both meteorology and biology.

Attempts have been made to establish limits on the variation of γ by studying the effects that such variation produces on the interior of the earth (see Murphy and Dicke (1964)). If Hypothesis B were true, these effects would be large and in principle observable. However, just as in the case of climatic effects, the geophysical effects of varying γ will be modified by a great variety of unknown factors resulting from the detailed thermal and chemical structure of the Earth's mantle. Until geophysics becomes an exact science, it will be impossible to decide with certainty whether the contribution to the heat-flow through the earth's crust produced by a decrease in γ should be positive or negative.

Paradoxically, we possess far more detailed and reliable knowledge of the structure and evolution of the sun than we do of the earth. It is therefore possible to analyze the effects of varying γ on the Sun's history, more precisely than we can analyze the terrestrial effects. Such an analysis of solar evolution was carried out by Pochoda and Schwarzs-child (1964). They did not use simple approximations such as (4.11) and (4.18), but computed in detail the evolution of solar models with varying γ, taking account of the real structural complications such as the gradual conversion of hydrogen to helium in the core. They found that the dependence of luminosity on γ is not quite as strong as (4.18) but is still very steep.

The conclusions of Pochoda and Schwarzschild are as follows. If the present age of the universe is 1.3×10^{10} years or less, then Dirac's Hypothesis B is excluded. In this case, if $\gamma \sim t^{-1}$, the sun cannot have existed as a main sequence star for 4.5×10^9 years and arrive now at its observed luminosity. Either it would have had too low an initial hydrogen abundance to last for 4.5×10^9 years, or it would have a high enough initial hydrogen abundance to last but would be reduced now to less than $\frac{3}{4}$ of its actual luminosity. However, if the age of the universe is 1.5×10^{10} years or more, there exists an initial hydrogen abundance which leads to the observed luminosity at the correct age of the sun.

At the time when Pochoda and Schwarzschild wrote, the preferred value of the Hubble constant H was 2.5×10^{-18} sec^{-1}, which made an age of 1.5×10^{10} years for the universe seem rather long. Now that the preferred H (Sandage (1971)) is reduced to 1.6×10^{-18} sec^{-1}, an age of 1.5×10^{10} years or greater is entirely acceptable. The evidence from solar evolution is therefore unable to decide for or against Dirac's hypothesis. A fortiori, the evidence cannot exclude the Brans–Dicke Hypothesis C (see Dicke (1966)).

There are many other astrophysical consequences of varying γ which might be subject to observational test. For example, if Dirac's hypothesis is true, there cannot exist stars as luminous as the sun and substantially older than the sun. There are in fact many stars of solar luminosity, in globular clusters and elsewhere, which are believed to be twice as old as the sun. However, the estimates of the ages of these stars are derived from evolutionary calculations which assume γ constant. We do not yet have a reliable way to calculate stellar ages independently of the behavior of γ, except in the case of the sun for which the earth and the meteorites provide independent evidence.

Another consequence of varying γ would be a gradual shift toward larger masses of the range of possible stable neutron-star models. Ancient neutron-stars would necessarily have low masses, and an old neutron-star might one day find itself suddenly outside the range of stability. If this happened, the star would presumably erupt, either into a white-dwarf configuration or into a cloud of interstellar gas. It is not clear whether the eruption should be expected to be gradual or explosive. If it were explosive, there should be an observable event looking like a nova or supernova with unusual characteristics.

5. DICKE'S COUNTER TO THE NUMEROLOGICAL ARGUMENT

5.1 STATEMENT OF DICKE'S ARGUMENT

First Dicke (1961), and then in a more elaborate fashion Carter (1970), raised a basic objection to the numerological argument of Dirac. Dirac's argument is that the two small ratios γ and δ cannot be equal by accident, and since δ varies with t^{-1} we should expect γ also to vary with t^{-1}. Dicke's objection is the following. Suppose that γ is constant. We can estimate the time t_D that the universe will take to develop stars, planets, living organisms, people and finally cosmologists who can speculate about these matters. Let then t_D be the time taken to produce a Dirac or a Dicke. Dicke's argument is that t_D must be comparable with the life-time of a main-sequence star. We do not know enough about biology to calculate t_D in detail. But a lower limit to t_D is set by the minimum life-time t_M of a main-sequence star, and it is reasonable to take t_M as determining roughly the order of magnitude of t_D.

The order of magnitude of t_M is given by

$$t_M = (\eta Nmc^2/L), \tag{5.1}$$

where

$$\eta = 0.008 \tag{5.2}$$

is the fraction, of the rest-energy Nmc^2 of a star, that is released in nuclear reactions when hydrogen is burnt. N is the number of protons in the star, and L is the luminosity. Following Carter (1970), we estimate N and L in the following way. We are concerned with stars on the upper main-sequence which have the largest values of (L/N) and consequently the shortest life-times.

5.2 MASS OF UPPER-MAIN-SEQUENCE STARS

The lower limit on the life-time of such stars arises from the fact that they become pulsationally unstable when the radiation pressure greatly exceeds the material pressure. Therefore an upper main-sequence star near the limit of stability will have radiation pressure comparable with material pressure. Apart from numerical factors, the radiation pressure is

$$p_R = (kT)^4(\hbar c)^{-3}, \tag{5.3}$$

where k is Boltzmann's constant and T is an average temperature in the star's interior. The material pressure is similarly of the order

$$p_M = Nr^{-3}kT, \qquad (5.4)$$

where r is the stellar radius. Equating these two pressures gives

$$rkT = N^{\frac{1}{3}}\hbar c. \qquad (5.5)$$

But the virial theorem also gives a relation between the mean thermal energy kT per particle and the mean gravitational potential, namely

$$kT = GNm^2r^{-1}. \qquad (5.6)$$

Thus kT and r both disappear when (5.5) and (5.6) are combined, giving the result

$$N = (Gm^2/\hbar c)^{-3/2} = \gamma^{-3/2}. \qquad (5.7)$$

So N, the maximum number of protons which can assemble to give birth to a stable main-sequence star, is determined by γ alone. Actually, when the correct numerical factors are put in, the maximum N is close to $10\gamma^{-3/2}$.

Parenthetically, it is interesting to observe that the value (5.7) of N is equal to the Chandrasekhar limit, the maximum number of protons (or neutrons) that can exist in a stable cold degenerate star. For a degenerate configuration, (5.5) is replaced by the law of Fermi statistics (the uncertainty principle)

$$rp_f = N^{\frac{1}{3}}\hbar, \qquad (5.8)$$

where p_f is the Fermi momentum of the degenerate particles. The virial theorem (5.6) is replaced by

$$p_f v_f = GNm^2r^{-1}, \qquad (5.9)$$

where v_f is the velocity of the particles on the Fermi surface. When (5.8) and (5.9) are combined, p_f and r disappear as before, and we obtain

$$N = (v_f/c)^{3/2}\gamma^{-3/2} < \gamma^{-3/2}. \qquad (5.10)$$

Remarkably, the mass-limit for the birth of stable stars is, apart from a numerical factor of the order of 10, the same as the mass-limit for stars which can enjoy a quiet death. This explains what would otherwise be a mystery, the fact that stars are mostly born with masses of the same order as the Chandrasekhar limit, as if they had foreknowledge of the conditions which are destined to determine their ultimate fate millions or billions of years later.

5.3 LUMINOSITY OF UPPER-MAIN-SEQUENCE STARS

For an upper-main-sequence star satisfying (5.5), the luminosity is roughly

$$L = (NkT/\tau), \tag{5.11}$$

where τ is the photon diffusion time given by

$$\tau = (Nm\kappa/rc). \tag{5.12}$$

The opacity κ for highly luminous stars is not given by the Kramers formula (4.11) but is determined by the Thompson scattering of photons by free electrons. Therefore

$$\kappa = \sigma/m, \tag{5.13}$$

where σ is the Thompson cross-section

$$\sigma = \alpha^2 \zeta^2 (\hbar/mc)^2, \tag{5.14}$$

and

$$\zeta = (m/m_e) = 1836 \tag{5.15}$$

is the ratio of proton to electron mass. Together, (5.6), (5.11) and (5.12) give

$$L = (NGm^2c/\sigma). \tag{5.16}$$

The life-time (5.1) then becomes

$$t_M = \eta \alpha^2 \zeta^2 \gamma^{-1} (\hbar/mc^2). \tag{5.17}$$

It happens that

$$\eta \alpha^2 \zeta^2 = 1.4 \tag{5.18}$$

is close to unity, and therefore (5.17) may be written

$$t_M = \gamma^{-1} (\hbar/mc^2). \tag{5.19}$$

5.4 CONCLUSION OF DICKE'S ARGUMENT

At a time t_M after the beginning of the universe, the value of the ratio $\delta = (H\hbar/mc^2)$ would be approximately

$$\delta_M = (\hbar/mc^2)t_M^{-1} = \gamma. \tag{5.20}$$

So the observed 'coincidence' that γ and δ are of the same order of smallness is explained by the fact that it takes roughly a time δ_M to produce observers whose existence is coupled to the existence of main-sequence stars. The orders of magnitude of δ_M and γ are bound to be the same, whether γ is varying with time or not. Thus Dirac's numerological argument for varying γ is destroyed, if we believe that the epoch t_D at which we are observing the universe is causally related to t_M.

Actually the equality between γ and δ is not so close as (5.20) would imply. The observed values differ substantially, namely

$$\delta \approx 2 \times 10^{-4} \gamma, \tag{5.21}$$

which implies that

$$t_D \approx 5000 t_M. \tag{5.22}$$

In fact, t_D is governed by the life-time of the sun rather than by the life-time of a massive upper-main-sequence star. Thus t_D, the life-time of the sun, is about 10^{10} years, whereas t_M, the minimum main-sequence life-time, is about 2×10^6 years. If we accept the view that the creation of cosmologists requires not merely a main-sequence star but a slowly-rotating G-type main-sequence star similar to the sun, then Dicke's argument explains not only the crude equality of δ and γ but also the magnitude of their inequality given by (5.21), (5.22). Carter (1970) has carried speculative arguments of this sort a good deal further, and has reached the conclusion that the presence in the universe of conscious observers places limits on the absolute magnitudes of γ and δ and not only on their ratio. But these further applications of Carter's 'principle of cognizability' are not required for the discussion of the question of the time-variation of γ.

6. SUMMING UP

Of the five hypotheses A, B, C, D, E with which this discussion began, only D and E, the two which included a time-variation of the fine-structure constant, are yet excluded by observation. Gamow's Hypothesis D is excluded by direct astronomical measurement of α, while Teller's Hypothesis E is excluded by the more indirect evidence of geophysical isotope abundances. The two hypotheses, B and C, which include a time-variation of the gravitational constant, are consistent with present observational knowledge. Dirac's Hypothesis B is only barely consistent with the evidence from solar and stellar evolution, and its tenability will be decisively tested by interplanetary ranging measurements within the next few years. There remain Dicke's Hypothesis C, with a much weaker time-variation of gravity than B, and the orthodox Hypothesis A in which all laboratory-measured constants are truly constant. A direct decision between A and C is not within the range of present observational capabilities. Probably the decision will be reached indirectly, by observations which do not look for variation of gravity but look for other phenomena for which the predictions of Brans–Dicke cosmology and orthodox general relativity differ.

236   F. J. DYSON

Altogether, we can expect that at most one of our five hypotheses will survive the experimental tests which will challenge them in the next ten years. It is quite possible that all five will fail, and then it will be up to the speculative cosmologists, and up to Dirac in particular, to think of something new.

REFERENCES

Bahcall, J. N. (1961). *Phys. Rev.* **124**, 495.
Bahcall, J. N. and Schmidt, M. (1967). *Phys. Rev. Letters* **19**, 1294.
Berényi, D. (1968). *Rev. Mod. Phys.* **40**, 390.
Brans, C. and Dicke, R. H. (1961). *Phys. Rev.* **124**, 925.
Brodzinski, R. L. and Conway, D. C. (1965). *Phys. Rev.* **138**, B1368.
Carter, B. (1970). 'Large numbers in astrophysics and cosmology', talk at Clifford Centennial meeting, Princeton, N.J. Princeton University preprint.
DeWitt, B. S. (1964). *Phys. Rev. Letters* **13**, 114.
Dicke, R. H. (1961). *Nature* **192**, 440.
Dicke, R. H. (1966). 'Stellar evolution with varying G', in *Stellar Evolution*, ed. R. F. Stein and A. G. W. Cameron (Plenum Press, N.Y.).
Dirac, P. A. M. (1937). *Nature* **139**, 323.
Dirac, P. A. M. (1938). *Proc. Roy. Soc. (London)* A**165**, 199.
Drever, R. W. P. and Payne, J. A. (1968). Private communication.
Gamow, G. (1967). *Phys. Rev. Letters* **19**, 759.
Gilbert, N. (1958). *Comptes Rendus* **247**, 868.
Gold, R. (1968). *Phys. Rev. Letters* **20**, 219.
Henderson, G. H. (1934). *Proc. Roy. Soc. (London)* A**145**, 591.
Herr, W., Hoffmeister, W., Hirt, B., Geiss, J. and Houtermans, F. G. (1961). *Z. Naturforsch.* **16a**, 1053.
Hirt, B., Tilton, G., Herr, W. and Hoffmeister, W. (1963). In *Earth Science and Meteoritics*, ed. J. Geiss and E. D. Goldberg (North-Holland, Amsterdam).
Isham, C. J., Salam. A. and Strathdee, J. (1971). *Phys. Rev.* D3, 1805.
Konopinski, E. J. (1943). *Rev. Mod. Phys.* **15**, 209.
Landau, L. D. (1955). 'On the quantum theory of fields', in *Niels Bohr and the Development of Physics*, ed. W. Pauli (Pergamon Press, London).
Murphy, C. T. and Dicke, R. H. (1964). *Proc. Am. Phil. Soc.* **108**, 224.
Peebles, P. J. and Dicke, R. H. (1962). *Phys. Rev.* **128**, 2006.
Pochoda, P. and Schwarzschild, M. (1964). *Astrophys. J.* **139**, 587.
Sandage, A. R. (1971). Private communication from Dr J. N. Bahcall.
Sciama, D. W. (1953). *Mon. Not. Roy. Ast. Soc.* **113**, 34.
Shapiro, I. I., Smith, W. B., Ash, M. B., Ingalls, R. P. and Pettengill, G. H. (1971). *Phys. Rev. Letters* **26**, 27.
Teller, E. (1948). *Phys. Rev.* **73**, 801.
Wilkinson, D. H. (1958). *Phil. Mag.* **3**, 582.

14

On the Time–Energy Uncertainty Relation

Eugene P. Wigner

1. INTRODUCTION AND SUMMARY

There is only one well-known application for the time–energy uncertainty relation: the connection between the life-time and the energy-width of resonance states. The relation in question was commonly known even before quantum mechanics was established; its first quantum mechanical derivation was based on Dirac's original theory of the interaction between matter and radiation.[1] The point which should be noted is that the uncertainty relation does not apply to time and energy *in abstracto* but to the life-time of a definite state of a system. In the example referred to, this is the state in which an atom is in an excited state but there is no radiation present.

The preceding formulation of the time–energy uncertainty relation appears to be different from Heisenberg's well-known position–momentum uncertainty relation.[2] This postulates that the state of any quantum mechanical system is, at every instant of time, such that the product of the position and momentum spreads s_x and s_p

$$s_x s_p \geqslant \tfrac{1}{2}\hbar. \tag{1}$$

The spreads in question are defined as the positive square roots of

$$s_x{}^2 = \langle \psi | (x - x_0)^2 | \psi \rangle / \langle \psi | \psi \rangle$$

$$s_p{}^2 = \langle \psi | (p - p_0)^2 | \psi \rangle / \langle \psi | \psi \rangle \tag{1a}$$

and holds for every ψ and all x_0, p_0. In (1a) ψ is an arbitrary state vector, x and p position and momentum operators, respectively. The clause 'at every instant of time' in the preceding formulation of the uncertainty principle is, evidently, important since the state vector ψ in (1) represents the state of the system under consideration only for a definite instant of time and this must be the same for both expressions in (1a).

It follows that, if one wishes to formulate the time–energy analogue of the usual position–momentum uncertainty relation, one will have to restrict one's attention to the situation along a single value of a space coordinate (which will be chosen as the x coordinate) just as a single time coordinate entered (1). The simplest form of the relation then becomes the statement that the product of the energy spread and the spread in the probability that the definite value of the coordinate x be assumed at time t, is at least $\hbar/2$. For a single non-relativistic particle these spreads τ and ϵ are defined as the positive square roots of

$$\tau^2 = \frac{\iiint |\psi(x, y, z, t)|^2 (t-t_0)^2 \, dy \, dz \, dt}{\iiint |\psi(x, y, z, t)|^2 \, dy \, dz \, dt} \tag{2}$$

$$\epsilon^2 = \langle \psi | (H - E_0)^2 | \psi \rangle / \langle \psi | \psi \rangle$$

or

$$\epsilon^2 = \frac{\iiint |\phi(x, y, z, E)|^2 (E - E_0)^2 \, dy \, dz \, dE}{\iiint |\phi(x, y, z, E)|^2 \, dy \, dz \, dE}, \tag{2a}$$

where

$$\phi(x, y, z, E) = \int \psi(x, y, z, t) e^{iEt/\hbar} \, dt \tag{2b}$$

is the Fourier transform of ψ from time into energy. It will be seen (as is pretty evident) that the uncertainty relation

$$\tau \epsilon > \tfrac{1}{2}\hbar \tag{3}$$

holds again for all values of t_0 and E_0. However, whereas there are, for any x_0 and p_0 in (1), state vectors for which the equality sign is valid in (1), namely those for which the x dependence of ψ is a Gaussian of $x - x_0$ multiplied with $\exp ip_0 x$, the inequality sign always holds in (3). This is a consequence of the fact that the energy is a positive definite operator (or has, in the non-relativistic case, a lower bound). The lower limit of $\tau \epsilon$ is, naturally, independent of t_0 but does depend on E_0 and increases substantially as E_0 approaches the lower bound of H from above. Naturally, it increases even further as E_0 crosses that bound and decreases further. All this, as well as equation (2a), will be further discussed below.

2. A GENERAL OBSERVATION ON THE INTERCHANGE OF THE SPACE AND TIME COORDINATES

Actually, the point just mentioned does not constitute the most interesting difference between the position–momentum and the time–energy uncertainty relations. The difference which appears most interesting to this writer will be first formulated in a universe with only one spatial dimension. In such a universe, it is natural to ask[3] for the probability that, at a definite time, the spatial coordinate of the particle have the value x. It is less natural to ask for the probability that the particle be, at a definite landmark in space, just at the time t. It would be more natural to ask, instead, for the probability that the particle crosses the aforementioned landmark at the time t from the left, and also that it crosses that landmark, at a given time, from the right. The sum of these probabilities, when integrated over all times t, is 1 for a free particle. The difference between the two cases arises from the fact that a particle's world line can cross the $t =$ constant line only in one direction (in the direction of increasing t); it can cross the $x =$ constant line in both directions. If we replace 'line' in the last sentence by 'plane', we have the generalization of the distinction to the actual four-dimensional universe. In Dirac's theory of the electron,[4] the probability that the particle be, at the definite time t, at the point x, y, z is given by $\psi^\dagger \alpha_0 \psi$, the space–time variables of ψ being x, y, z, and t. The probabilities for crossing the $x =$ constant plane, at y, z and at time t, in the two different directions, are given, essentially, by $\frac{1}{2}\psi^\dagger(\alpha_0+\alpha_x)\psi$ and $\frac{1}{2}\psi^\dagger(\alpha_0-\alpha_x)\psi$. This point of this paragraph, interesting though it may be, will not be elaborated further.

Even though the idea of space–time, and hence the similarity between space and time coordinates, appears natural from the point of view of relativity theory, initial conditions characterizing the state of the system at a definite instant of time appear more natural to us than initial conditions giving the state of the system for all times but only for a single value of one of the spatial coordinates. There are, indeed, valid reasons for this preference. The transition from the position–momentum uncertainty relation to the time–energy uncertainty relation is, however, based on the second type of description of the state of the system. It would be interesting to develop equations of motion giving the spatial derivative of the second type of characterization of a state and to explore the properties of the resulting equations.

3. A GENERALIZATION OF HEISENBERG'S UNCERTAINTY RELATION

There are uncertainty relations for practically any two non-commuting operators, but the generalization of Heisenberg's relation to be pointed out here is a very special one and bears a close resemblance to the original form of the relation. We denote, first, the variables of ψ by

x and r, the latter one standing for all variables different from x. The relation (1) then remains valid if one replaces ψ in (1a) by

$$\psi(x, r) \rightarrow \int \phi(r)^* \psi(x, r) dr \tag{4}$$

ϕ being any function of r and the integration is over all values of the continuous coordinates implied by r and summation over the discrete ones. This is, of course, a well-known fact, most commonly used with a ϕ which is a delta function of all coordinates different from x. The derivation of the relation (1), given by Robertson,[2] remains equally valid, however, if r is assumed to involve also the time, with $\psi(x, r)$ satisfying Schrödinger's equation and ϕ depending on time in an arbitrary fashion. The right side of (4) is then a generalized transition amplitude which can be thought of as corresponding to a measurement lasting a finite length of time, but not affecting x.

Accepting the generalization of Heisenberg's relation just outlined, one sees that the time–energy uncertainty relation referring to the life-time of resonance states, which was mentioned in the first paragraph of this article, is not as different from (3) as it first appeared. It is in fact included in the generalization of (3) which is the analogue of the generalization of (1) just pointed to. The generalization replaces in this case $\psi(x, y, z, t)$ by $\langle u|\psi(t)\rangle = \chi(t)$ where u is any state vector. The time spread τ is then the positive square root of

$$\tau^2 = \frac{\int |\langle u|\psi(t)\rangle|^2 (t - t_0)^2 \, dt}{\int |\langle u|\psi(t)\rangle|^2 \, dt} = \frac{\int |\chi(t)|^2 (t - t_0)^2 \, dt}{\int |\chi(t)|^2 \, dt}. \tag{5}$$

In order to define the energy spread, we calculate, in analogy to (2b)

$$\phi(E) = (2\pi\hbar)^{-\frac{1}{2}} \int \psi(t) e^{iEt/\hbar} dt. \tag{5a}$$

This is a stationary state of energy E. Hence, ϵ^2 will be

$$\epsilon^2 = \frac{\int |\langle u|\phi(E)\rangle|^2 (E - E_0)^2 \, dE}{\int |\langle u|\phi(E)\rangle|^2 \, dE} = \frac{\int |\eta(E)|^2 (E - E_0)^2 \, dE}{\int |\eta(E)|^2 \, dE} \tag{5b}$$

where

$$\eta(E) = \langle u|\phi(E)\rangle = (2\pi\hbar)^{-\frac{1}{2}} \int \langle u|\psi(t)\rangle e^{iEt/\hbar} dt$$
$$= (2\pi\hbar)^{-\frac{1}{2}} \int \chi(t) e^{iEt/\hbar} dt \tag{5c}$$

is the Fourier transform of χ. The factor $(2\pi\hbar)^{-\frac{1}{2}}$ renders the denominators of (5b) and (5) equal. With these definitions, and identifying u with the state vector of the resonance, (3) will represent the uncertainty relation giving the minimum energy spread of a resonance with given life-time. This is the time–energy uncertainty relation mentioned in the first paragraph of the present article. On the other hand, if one inserts a delta function of x for u, the close analogue of the original Heisenberg uncertainty relations given by (2), (2a), and (3) results.

We will go over next to the proof of (3) with the τ and ϵ given by (5) and (5b) and will give some estimates for the lowest possible value of $\tau\epsilon$ as function of E_0. The lower bound of $\tau\epsilon$ will turn out, naturally, to be independent of t_0 and, perhaps somewhat less obviously, depends only on the ratio E_0/ϵ, not on E_0 and ϵ separately.

4. MINIMAL TIME–ENERGY UNCERTAINTY
PRELIMINARY REMARKS

As was mentioned already in the first section, the product of the spread in the time of presence at a definite plane and the spread in energy, $\tau\epsilon$, has to exceed $\hbar/2$. (The time of presence at a definite plane was, actually, generalized in the preceding section to the time of being in any definite quantum mechanical state.) The rest of this article will be concerned with the lower bound of $\tau\epsilon$, for a given ϵ, by characterizing the ψ which renders $\tau\epsilon$ to a minimum. This ψ, and the corresponding $\tau\epsilon$, will in this case depend on the energy E_0 around which the spread of the energy of ψ is defined as ϵ. The minimum of the quantity $\tau\epsilon$ will be independent of t_0 around which the spread of the time of presence, that is τ, will be calculated, ψ will depend on t_0 only in a trivial way. The last two statements follow, of course, from time displacement invariance. The minimum of $\tau\epsilon$ (and the corresponding ψ) will, on the other hand, depend on the choice of ϵ: clearly, if ϵ can be chosen to be very small as compared with the excess of E_0 over the lower bound of the energy, the existence of the lower bound will have very little significance. The lower bound $\hbar/2$ of $s_x s_p$ of the usual Heisenberg uncertainty relation was independent of s_p because the momentum p has no lower bound. In our case, the lower bound of $\tau\epsilon$ can be expected to increase with increasing ϵ (or, rather, with the increase of the ratio $\epsilon/(E_0 - E_b)$ where E_b is the lower bound of the energy).

Actually, what the subsequent calculations are aimed at is the determination of

$$\chi(t) = \langle u|\psi(t)\rangle \tag{6}$$

or of its Fourier transform

$$\eta(E) = \langle u|\phi(E)\rangle = (2\pi\hbar)^{-\frac{1}{2}} \int \chi(t) e^{iEt/\hbar} dt \tag{6a}$$

which render

$$\tau^2\epsilon^2 = \frac{\int |\chi(t)|^2 (t - t_0)^2 \, dt \int |\eta(E)|^2 (E - E_0)^2 \, dE}{(\int |\eta(E)|^2 \, dE)^2} \tag{7}$$

to a minimum. In the calculation which follows $\eta(E)$ will be the dependent variable because the essential limitation of the problem, the existence of a lower bound for the energy, is most easily expressed in terms of $\eta(E)$. This must vanish for all E below the lower bound. It will be convenient to choose an energy scale in which the lower bound is 0; the transformation to any other lower bound is trivial. As a result, the integrations over E will extend from 0 to ∞ and will not be specified explicitly.

We wish to establish, next, that for $0 \leq E$ the function $\eta(E)$ can be chosen arbitrarily, subject only to the condition that it approach 0 fast enough as E increases toward infinity. This requires three assumptions. First, the Hamiltonian will be assumed to have a continuous spectrum, extending from its lower bound to infinity. This is surely true for any isolated system. Second, it will be assumed that, if $|u\rangle$ is expanded in terms of the characteristic functions $|E\rangle$ of the Hamiltonian

$$|u\rangle = \int b(E)|E\rangle dE \tag{8}$$

$b(E)$ does not vanish for any positive E. This condition is surely fulfilled for all states $|u\rangle$ which restrict the system to a spatial plane, i.e., for the original form (3) of the time–energy uncertainty relation. It is fulfilled also for the resonance states discussed later but is, of course, not true for all u. The second assumption, therefore, restricts our considerations to a certain degree. The third assumption is that the $b(E)$ of (8) does not go to 0 at very large E as fast as $\exp(-\alpha E^2)$ with any α. This is again fulfilled in the two aforementioned cases but restricts u somewhat further.

If the preceding conditions are satisfied, we expand $\psi(t)$ also in terms of the characteristic functions of the Hamiltonian

$$\psi(t) = \int a(E)e^{-iEt/\hbar}|E\rangle dE. \tag{9}$$

If the spectrum of the Hamiltonian is degenerate, the $|E\rangle$ in (9) shall be the same which appear in the expansion of $|u\rangle$ in (8). We then have, by (6)

$$\chi(t) = \int \int b(E)^* \langle E|E'\rangle a(E')e^{-iEt/\hbar} dE dE'$$
$$= \int b(E)^* a(E)e^{-iEt/\hbar} dE \tag{10}$$

and hence

$$\eta(E) = (2\pi\hbar)^{\frac{1}{2}} b(E)^* a(E). \tag{10a}$$

Hence, any $\eta(E)$ can be obtained by a proper choice of $a(E)$ as long as $b(E)$ remains different from zero for all E. The square integrability of $a(E)$ does, though, impose a condition on the way η must go to 0 as

$E \to \infty$. It will turn out, however, that the η which will be obtained under the sole condition that it be square integrable goes to 0 as $\exp\left(-\frac{1}{2}cE^2\right)$. Hence, if u satisfies the last condition imposed above, the $a(E)$ needed to furnish this η will be automatically square integrable. We can, therefore, proceed to the determination of the $\eta(E)$ which gives the smallest τ_ϵ, permitting η to be, for positive E, an arbitrary function of E.

5. MINIMAL TIME-ENERGY UNCERTAINTY. EQUATION FOR η WHICH MINIMIZES UNCERTAINTY

If $\eta(E)$ were finite at the lower bound of the energy (at $E=0$), or if it had a discontinuity somewhere, its Fourier transform

$$\chi(t) = (2\pi\hbar)^{-\frac{1}{2}} \int \eta(E)e^{-iEt/\hbar} \, dE$$

would go to 0 at $t = \infty$ only as $1/t$. The integral in (5) then would become infinite. This surely would not give the minimum value of τ_ϵ. It follows that $\eta(0)=0$. Hence

$$\tau^2 = \frac{\int_{-\infty}^{\infty} dt(t-t_0)^2 \int\eta(E)e^{-iEt/\hbar} \, dE \int\eta(E')^* e^{iE't/\hbar} \, dE'}{2\pi \hbar \int |\eta(E)|^2 \, dE}. \tag{11}$$

Introducing $\xi = (t-t_0)/\hbar$ as variable instead of t, and writing

$$\eta_0(E) = \eta(E)e^{-iEt_0/\hbar} \tag{12}$$

one obtains

$$\tau^2 = \frac{\hbar^2 \int_{-\infty}^{\infty} d\xi \, \xi^2 \int\eta_0(E)e^{-iE\xi} \, dE \int\eta_0(E')^* e^{iE'\xi} dE'}{2\pi \int |\eta_0(E)|^2 \, dE}$$

$$= \frac{\hbar^2}{2\pi} \frac{\int_{-\infty}^{\infty} d\xi \int\int\eta_0(E) \, \eta_0(E')^* (\partial^2/\partial E \partial E') e^{i(E'-E)\xi} \, dEdE'}{\int |\eta_0(E)|^2 \, dE}. \tag{13}$$

Since $\eta_0(E)$ and $\eta_0(E')^*$ both vanish at both ends of the integration, at 0 and ∞, partial integration with respect to E and E' gives

$$\tau^2 = \frac{\hbar^2}{2\pi} \frac{\int_{-\infty}^{\infty} d\xi \int\int(\partial\eta_0(E)/\partial E)(\partial\eta_0(E')^*/\partial E')e^{i(E'-E)\xi} \, dEdE'}{\int |\eta_0(E)|^2 \, dE}$$

$$= \hbar^2 \int |\partial\eta_0(E)/\partial E|^2 \, dE/ \int |\eta_0(E)|^2 \, dE. \tag{14}$$

The last line follows from Fourier's theorem and is the expression which had to be expected. In fact, the calculation was carried out in such detail principally to show the necessity to assume $\eta(0) = 0$ in order to obtain (14).

We now have

$$\tau^2 = \hbar^2 \int |\partial \eta_0(E)/\partial E|^2 dE / \int |\eta_0(E)|^2 dE \qquad (15)$$

and

$$\epsilon^2 = \int |\eta_0(E)|^2 (E - E_0)^2 dE / \int |\eta_0(E)|^2 dE. \qquad (15a)$$

If one writes

$$\eta_0(E) = \alpha(E) e^{i\beta(E)} \qquad (16)$$

with both α and β real, β drops out from the expression for ϵ^2 and the denominator of (15). The integral in the numerator becomes

$$\int |\partial \alpha/\partial E + i\alpha \partial \beta/\partial E|^2 dE = \int (\partial \alpha/\partial E)^2 + \alpha^2 (\partial \beta/\partial E)^2 dE \qquad (17)$$

and this will be decreased if one sets $\partial \beta/\partial E = 0$. It follows that the minimum of $\tau \epsilon$ will be assumed for a real η_0 and such an η_0 will be assumed henceforth. The real nature of η_0 could have been inferred also from time inversion invariance.

We are ready to obtain the minimum of $\tau^2 \epsilon^2$ for given ϵ^2 (and E_0). To obtain it, we set the variation of

$$\tau^2 \epsilon^2 + \lambda' \epsilon^2 = \hbar^2 \epsilon^2 \frac{\int (\partial \eta_0/\partial E)^2 \, dE}{\int \eta_0{}^2 \, dE} + \lambda' \frac{\int \eta_0{}^2 (E - E_0)^2 \, dE}{\int \eta_0{}^2 \, dE} \qquad (18)$$

equal to 0; the λ' is a Lagrange multiplier. This gives us the equation

$$\int \eta_0{}^2 \, dE \left[-2\hbar^2 \epsilon^2 \frac{\partial^2 \eta_0}{\partial E^2} + 2\lambda' (E - E_0)^2 \eta_0 \right]$$

$$= [\hbar^2 \epsilon^2 \int (\partial \eta_0/\partial E)^2 \, dE + \lambda' \int \eta_0{}^2 (E - E_0)^2 \, dE] 2\eta_0. \qquad (19)$$

Elimination of the integrals by means of (15) and (15a) and division by $2\tau^2 \epsilon^2$ gives then

$$-\frac{\hbar^2}{\tau^2} \frac{\partial^2 \eta_0}{\partial E^2} + \frac{\lambda}{\epsilon^2} (E - E_0)^2 \eta_0 = (1 + \lambda)\eta_0 \qquad (20)$$

where

$$\lambda = \lambda'/\tau^2 \qquad (20a)$$

has been introduced to make (20) somewhat more symmetric; λ must be so determined that (15a) become valid.

Since η_0 must vanish for both $E = 0$ and $E = \infty$, (20) is essentially a characteristic value – characteristic function equation. It is well known, and can be easily verified, that its solution which approaches 0 for very large E approaches 0 as $\exp(-\tfrac{1}{2}cE^2)$ with $c = \lambda^{\frac{1}{2}}\tau/\hbar\epsilon$. This verifies the statement made about η_0 at the end of the last section.

The solution of (20) is easily obtained for very large E_0/ϵ and for $E_0 = 0$. In the former case, one can set $\lambda = 1$ and obtains $\tau = \hbar/2\epsilon$ (that is $\tau\epsilon = \hbar/2$ as expected) and

$$\eta_0 = \exp(-(E-E_0)^2/4\epsilon^2). \tag{21}$$

This does not quite satisfy the boundary condition at $E = 0$ but for large E_0 it satisfies it quite closely. Similarly, (15a) is satisfied closely enough. In the present case, actually, the minimum of $\tau\epsilon$ is independent of ϵ as long as this remains very much smaller than E_0.

The other case in which (20) can be easily solved is $E_0 = 0$. In this case the only solutions of (20) which satisfy the boundary conditions have the form

$$\eta_0 = E \exp(-\tfrac{1}{2}cE^2). \tag{22}$$

One can determine c by (15a) to be

$$c = 3/2\epsilon^2. \tag{22a}$$

This then solves the differential equation (20) if one sets again $\lambda = 1$ and $\tau\epsilon$ becomes

$$\tau\epsilon = 3\hbar/2. \tag{22b}$$

Because of the low value of E_0, the uncertainty is much larger than in Heisenberg's relation or for the large E_0 of (21). In this case again, the minimum of $\tau\epsilon$ is independent of ϵ – it is, as we shall see next, independent in no other case but the two just considered. Let us then proceed to the determination of λ and the discussion of η_0 in the general case.

6. MINIMAL TIME–ENERGY UNCERTAINTY. DISCUSSION

Let us observe, first, that the lower limit of $\tau\epsilon$ depends on ϵ and E_0 only in terms of E_0/ϵ. The two examples considered in the last section correspond to the values of ∞ and 0, respectively, of this quantity. It is for this reason that we found the minimum of $\tau\epsilon$ to be independent of ϵ and E_0 separately.

In order to see that the minimum of $\tau\epsilon$ depends only on E_0/ϵ, let us denote the solution of (20) for $\epsilon = 1$ and the value e_0 of E_0 by $g(E, e_0)$. This solves (20) with $\epsilon = 1$, $E_0 = e_0$ and a $\lambda(e_0)$ such that the solution g satisfies the condition (15)

$$\int g(E, e_0)^2 (E-e_0)^2 dE = \int g(E, e_0)^2 dE \tag{23}$$

and the differential equation

$$-\frac{\hbar^2}{\tau^2}g''(E,\ e_0)+\lambda(E-e_0)^2g(E,\ e_0)=(1+\lambda)g(E,\ e_0), \qquad (23a)$$

the primes denoting differentiation with respect to the first variable and τ the value for which $g(0,\ e_0)=0$, and g tending to 0 for large E.

We can then set

$$\eta_0(E)=g(E/\epsilon,\ E_0/\epsilon) \qquad (24)$$

i.e., choose $e_0=E_0/\epsilon$ and expand the first variable of g by the factor ϵ. If one then substitutes (24) into (23a), one finds that η_0 satisfies (20). Similarly,

$$\int\eta_0(E)^2(E-E_0)^2dE=\ \int g(E/\epsilon,\ E_0/\epsilon)^2(E-E_0)^2dE$$

$$\epsilon^3\int g(E',\ E_0/\epsilon)^2(E'-E_0/\epsilon)^2dE'=\epsilon^2\int g(E/\epsilon,\ E_0/\epsilon)^2dE=\epsilon^2\int\eta_0(E)^2dE.$$

The before-last member is a consequence of (23). This establishes the theorem which was plausible anyway. It follows that the wave functions η_0 of the minimal $\tau\epsilon$, that is time–energy uncertainty, can all be obtained by means of (24) by solving (23a) with the subsidiary condition (23).

In order to solve (23a) with this subsidiary condition, one will choose a $\lambda\tau^2$, multiply (23a) with τ^2, whereupon this will be a simple characteristic value equation for τ^2. One obtains its lowest characteristic value and the corresponding characteristic function g by Ritz' or some other method and compare then the two sides of (23). If the left side is larger than the right, one will choose a larger $\lambda\tau^2$; if the left side is smaller, one will try a smaller $\lambda\tau^2$. It should be possible to satisfy (23) after not too many trials.

It may be of some interest to deduce a few identities between λ, τ and the function η_0 which renders $\tau\epsilon$ to a minimum. For this purpose, one multiplies (20) with certain factors and integrates the resulting equation with respect to E. Multiplication with η_0 and partial integration of the first term gives

$$(\hbar^2/\tau^2)\int(\partial\eta_0/\partial E)^2dE+(\lambda/\epsilon^2)\int(E-E_0)^2\eta_0^2dE=(1+\lambda)\int\eta_0^2dE. \qquad (25)$$

The integrated terms are 0 in this case because $\eta_0(0)=0$. Because of (15) and (15a), this is an identity, λ dropping out.

Multiplication of (20) with $\partial\eta_0/\partial E$ gives in a similar way

$$\frac{\hbar^2}{2\tau^2}\eta'_0(0)^2=\frac{\lambda}{\epsilon^2}\int\eta_0(E)^2(E-E_0)\ dE \qquad (25a)$$

showing that the mean value of the energy is always larger than E_0.

Finally, multiplication of (20) with $E\partial\eta_0/\partial E$ gives with (15) and (15a)

$$\frac{1}{2} - \frac{3}{2}\lambda - \frac{\lambda E_0 \int \eta_0(E)^2 (E-E_0)\, dE}{\epsilon^2 \int \eta_0(E)^2\, dE} = -\frac{1}{2} - \frac{1}{2}\lambda. \qquad (25b)$$

This last equation does not contain τ, the former one gives an expression for $\lambda\tau^2$ in terms of η_0. Naturally, (25a) and (25b) can be combined in various ways to eliminate λ or the integral in (25a). These equations play the role of the virial theorem which applies for the wave function giving minimal position-energy spread and Heisenberg's uncertainty relation. The last term on the left of (25b) vanishes in the simple cases considered in the last section: $E_0 = 0$ in the second case and the integral in the numerator vanishes for $E_0 = 0$, giving $\lambda = 1$ in both cases. In all other cases, $\lambda < 1$.

REFERENCES

1. P. A. M. Dirac, *Proc. Roy. Soc. (London)* **A114**, 243, 710 (1927). The calculation was carried out by V. Weisskopf and E. Wigner, *Z. Physik* **63**, 54 (1930). See also the article of the same authors, *ibid.* **65**, 18 (1930) and many subsequent discussions of the same subject and of resonance states decaying by the emission of particles rather than radiation.
2. W. Heisenberg, *Z. Physik* **43**, 172 (1927). For a rigorous derivation, see H. P. Robertson, *Phys. Rev.* **34**, 163 (1929). The derivation of section 5 is patterned on that of this article.
3. G. R. Allcock, *Ann. Phys.* (N.Y.) **53**, 253, 286, 311 (1969). Section II of the first of these articles gives a very illuminating discussion of the ideas which underlie also the present section. It also contains a review of the literature of the time–energy uncertainty principle, making it unnecessary to give such a review here. The review also gives a criticism of some of the unprecise interpretations of the time–energy uncertainty relation which are widely spread in the literature. The later parts of the aforementioned articles arrive at a pessimistic view on the possibility of incorporating into the present framework of quantum mechanics time measurements as described by Allcock in section II or in the present section. This pessimism, which is not shared by the present writer, is expressed, however, quite cautiously and is mitigated by the various assumptions on which it is based.
4. The notation of chapter XI of P. A. M. Dirac's *The Principles of Quantum Mechanics* (Oxford University Press, various editions) is used.

15

The Path-Integral Quantization of Gravity

Abdus Salam and J. Strathdee

1. INTRODUCTION

Together with Maxwell's and Dirac's equations, Einstein's equations for gravity represent high watermarks of human thought. Over the last decades much effort has been devoted to quantizing Einstein's equations[1] – and in this Dirac, as in the case of the other two equations, has been the pioneer. He has given a beautiful account of this work in his paper published in *Contemporary Physics*.[2]

Dirac starts by saying: 'There is no experimental evidence for the quantization of the gravitational field, but we believe quantization should apply to all the fields of physics. They all interact with one another and it is difficult to see how some could be quantized and others not.' Dirac's approach to the quantization problem is via the canonical formalism. We shall give a brief account of his method shortly; all we wish to remark here is that Dirac succeeds but, in his own view, not quite. There are in his final formulation, distribution-theoretic ambiguities of the type $\delta(\mathbf{x}) \otimes \delta(\mathbf{x})$.

Now in addition to the canonical formalism, there is an alternative formalism for quantization of field theories – the so-called sum-over-histories approach. Apart from its abstract virtues, this is the approach more favoured by S-matrix particle theorists, for it accords more readily with the immediacy of their need of writing down rules for computing scattering matrix elements. It is not commonly realized that it was Dirac himself who in the famous section 32 of his *Principles of Quantum Mechanics* first laid the foundations of the basic physical and mathematical ideas of this approach. It, of course, needed the genius of Feynman[3] to recognize the significance of these ideas and to build the approach into a formalism which has come, in the view of some physicists, even near to supplanting the canonical quantization scheme itself. This formalism was recently applied to gravity theory

T

by Faddeev and Popov. In this note we try to contrast the achievements and problems of this approach with the canonical formulation and in particular to trace the inter-relationships of the two methods. We feel that the path-integral method could possibly be extended and made mathematically more precise. If this could be achieved there may even be the possibility of resolving the difficulties Dirac encountered in the canonical quantization of gravity. Since the path-integral method originated with Dirac we wish to dedicate this note to him in the hope that he may feel persuaded to help with the achieving of the required precision of this method, particularly in connection with the quantization problem of Einstein's gravity.

2. A BRIEF REVIEW OF DIFFERENT APPROACHES TO THE QUANTIZATION OF GRAVITY

One of the main approaches to the quantization problem is via the canonical formalism and it was in the work of Dirac that the classical equations of Einstein were first given a canonical formulation.[4] For canonical coordinates Dirac chose the six spatial components, g_{rs} (r, $s = 1$, 2, 3), of the metric tensor on a spacelike surface, $x^0 = $ constant. The conjugate momenta, π^{rs}, are related to the Christoffel symbols Γ^0_{rs}. The complicating feature in this formalism is the existence of four constraint equations, $R_\mu(g, \pi) = 0$, whose consistency it is difficult to maintain in the quantized version. When g and π are replaced by operators which satisfy the canonical commutation rules, then the operators R_μ no longer commute with one another and it becomes a problem to decide whether there exist any states which simultaneously satisfy all the constraints, $R_\mu| \rangle = 0$. In Dirac's view it has not been possible to solve this problem in a convincing way because the local operators $R_\mu(g, \pi)$ are given in terms of products of the fields g_{rs} and π^{rs} at the same space–time point x^μ. Since the field operators are distributions rather than ordinary functions, such products are not well defined (principally because one does not know how to order the factors in R_μ, e.g. $\pi^2 g^2$ and $g^2 \pi^2$ differ by undefined terms of the type $(\delta_3(\mathbf{x}))^2$) and this means, in particular, that the very delicate consistency question cannot be analysed except in rather formal terms.[1]

The canonical approach was developed also by Arnowitt, Deser and Misner[5] who avoided the consistency problem by using the constraints $R_\mu(g, \pi) = 0$ to eliminate some of the variables g_{rs} and π^{rs} in the classical version. By using the freedom to impose four coordinate conditions as well, they were able to reduce the canonical variables to the two independent components $g_{rs}{}^{TT}$ (the transverse traceless part of g_{rs}) and their conjugates $(\pi^{rs})^{TT}$. By associating operators with only these canonically independent components the consistency problem is avoided. An unsatisfactory feature of this method lies in the fact that the Hamiltonian density cannot be exhibited as an explicit function of the independent canonical variables. This is because the (non-linear) constraint equations cannot be solved explicitly.

The canonical methods destroy manifest Poincaré covariance. That the theory is indeed covariant must be verified by exhibiting a set of generators of the Poincaré group and showing that they satisfy the correct commutation rules. In the Dirac approach this involves defining the constraint operators R_μ and showing that their commutators coincide with the classical Poisson bracket relations. There is an analogous problem in the approach of Arnowitt, Deser and Misner which involves the ordering of operators in the (implicit) Hamiltonian density.[1] In neither case can this question be resolved in a completely satisfactory way.

A formal analysis of the consistency problem has been given by Schwinger[6] who succeeded in defining the operators R_μ in such a way as to make the existence of states satisfying all the constraints at least plausible. The difficulty with Schwinger's work is that it involves manipulations with objects like $\delta(\mathbf{x})\delta(\mathbf{x})$ which are not well defined. Even if such quantities can be given an interpretation in the immediate context of the commutators, $[R_\mu, R_\mu]$, it is not clear that this interpretation will prove sufficient to make the theory well defined.

In addition to the approaches discussed above, a number of manifestly Poincaré covariant methods have been developed. These include the work of Gupta,[7] DeWitt[8] and Mandelstam.[9] There is also the approach of Bergmann which places greater stress on group-theoretic methods. Such methods have their respective advantages or disadvantages relative to each other and to the canonical methods. We shall not discuss them. They have in common the problem of factor ordering.

As stated in the introduction, we shall discuss here in more detail the Feynman path integral formalism for quantization.[10] Here the eigenvalue problem of conventional operator quantum mechanics is replaced by the problem of defining and evaluating an integral. The underlying idea of this method is to represent the Green's functions or transition amplitudes of a quantized theory by infinitely multiple integrals over an appropriate space of functions. Although these integrals have not been given a rigorous mathematical sense, they provide a flexible and comprehensive framework within which it is possible to discuss, on a formal level, both the canonical and the covariant formulations including such questions as the development of perturbation series with their associated Feynman rules, covariance, unitarity, gauge conditions and Ward–Takahashi identities.[11] The method as developed up to now, however, fails to resolve the difficulties of factor ordering and the associated quadratic products of δ-functions which presumably make their appearance in the form of undefined singular distributions of the type $(\partial^2(1/x^2))^2$ encountered when Feynman rules are written down for graphs in gravity theory.

There are other difficulties. To define the path integral for quantized general relativity it would be necessary, first of all, to fix the domain of integration. Over what space of functions $g_{\mu\nu}(x)$ does one integrate? Should the signature, topology and asymptotic behaviour of these metrics be specified in some way? In particular, would it be sufficient to confine the integral to metrics of the usual signature which have Euclidean topology and are asymptotically flat? Is there an asymptotic symmetry group (such as the Bondi–Metzner–Sachs group) which leaves the domain invariant? Secondly, to determine the factor ordering it would be necessary to define the integral as the limit of a multiple integral on a mesh of discrete space–time points as the mesh becomes dense. Does such a limit exist and is there some uniqueness (mesh independence) to it? With the answer to this question would be bound up the resolution of the factor-ordering dilemmas encountered by Dirac in the canonical approach. Finally, can the integral be regularized in such a way as to preserve a covariance group which is larger than the Poincaré group?

These questions are not very well posed. However, they may serve to indicate something of the nature of the problems which must be overcome if the path-integral approach is to lead to a satisfactory quantization of the gravitational field. Having remarked on the difficulties of the method, we shall now go on to consider some of its positive features.

3. THE PATH-INTEGRAL FORMULATION

The starting point in the path-integral quantization of Einstein's theory is the assumption that the Green's functions can be represented in the form

$$\langle Tg_{\kappa\lambda}(x_1)g_{\mu\nu}(x_2) \ldots \Gamma^\rho_{\sigma}{}^\tau(x_n)\rangle$$
$$= \int(dgd\Gamma)g_{\kappa\lambda}(x_1)g_{\mu\nu}(x_2)\ldots\Gamma^\rho_{\sigma\tau}(x_n)\exp((i/\hbar)\mathscr{S}(g,\,\Gamma)) \qquad (1)$$

where $\mathscr{S}(g, \Gamma)$ denotes the action functional integrated over all space-time, and $(dgd\Gamma)$ denotes a functional measure which is normalized to give

$$1 = \int(dgd\Gamma)\, \exp((i/\hbar)\mathscr{S}(g, \Gamma)). \tag{2}$$

The integral is supposed to extend in some sense over a space of functions $g_{\mu\nu}(x)$, $\Gamma^{\lambda}_{\mu\nu}(x)$. (As remarked before the precise definition of this integral is beyond our present resources, and for this reason all the considerations which follow have a formal character.)

The first problem to settle is the form of the action functional. The prescription usually adopted in path-integral quantization is to take for $\mathscr{S}$ the *classical* action functional. This turns out not to be sufficient in the case of relativity, however. It will in general be necessary to include additional terms in $\mathscr{S}$ which depend on Planck's constant $\hbar$. A method for constructing these terms has been given by Faddeev and Popov,[11] which we shall summarize. The idea is to write $\mathscr{S}$ as the sum of two pieces

$$\mathscr{S} = \mathscr{S}_P + \mathscr{S}_\Phi \tag{3}$$

where $\mathscr{S}_P$ denotes the classical Palatini action and $\mathscr{S}_\Phi$ denotes a supplementary term which serves to determine the gauge. The Palatini action is generally covariant and local. The supplementary term is not generally covariant and in most cases it is non-local. Moreover, it has a non-classical dependence on $\hbar$ in these cases. The classical ($\hbar$ independent) part of $\mathscr{S}_\Phi$ incorporates a set of four coordinate conditions which are needed to supplement the generally covariant equations of Einstein. Although in classical general relativity it is not usual to include a non-covariant term in the action but rather to impose the coordinate conditions independently, this term is essential in the quantized theory. This is because the integrals (1) and (2) would diverge if the action were constant on the orbits $\{g^\Omega\, \Gamma^\Omega\}$ generated by coordinate transformations. On the other hand, the classical requirement that physically interesting quantities should be independent of coordinate conditions must have an analogue in the quantized theory. It was shown by Faddeev and Popov that such a requirement is embodied in the constraint

$$1 = \int(d\Omega)\, \exp\left((i/\hbar)\int dx\, \mathscr{S}_\Phi(g^\Omega, \Gamma^\Omega)\right) \tag{4}$$

which restricts the functional form of $\mathscr{S}_\Phi$. The Faddeev–Popov (F–P) condition (4) is to be understood as an identity in the functions $g_{\mu\nu}(x)$ and $\Gamma^{\lambda}_{\mu\nu}(x)$. The functions $g^\Omega_{\mu\nu}$ and $\Gamma^{\lambda\Omega}_{\mu\nu}$ denote the respective transforms of $g_{\mu\nu}$ and $\Gamma^{\lambda}_{\mu\nu}$ which are generated by the coordinate transformation $\Omega : x^\mu \to x^\mu = \Omega^\mu(x)$. These transformations are characterized by four functions $\Omega^\mu(x)$ and the integral (4) is supposed to extend in some sense over the space of these functions on which $(d\Omega)$ denotes an invariant measure. This integral is at least as obscure as the original ones (1) and (2). At a formal level, however, it is necessary only to remember that the F–P condition (4), which constrains the functional form of $\mathscr{S}_\Phi$, is to be satisfied identically in g and Γ.

The role played by the F–P condition in quantized general relativity theory is best illustrated by examining the behaviour under changes of gauge of the quantum amplitude which corresponds to some classical functional $F(g, \Gamma)$. In the gauge determined by $\mathscr{S}_{\Phi_1}$ this quantum amplitude is given by

$$\langle TF(g, \Gamma)\rangle_{\Phi_1} = \int (dg d\Gamma) F(g, \Gamma) \exp\left((i/\hbar)(\mathscr{S}_P(g, \Gamma) + \mathscr{S}_{\Phi_1}(g, \Gamma))\right). \quad (5)$$

A similar integral defines the amplitude $\langle TF(g, \Gamma)\rangle_{\Phi_2}$. To obtain the transformation rule which connects the two amplitudes we shall assume that, in the integral (5), both the measure $(dg d\Gamma)$ and the domain are invariant under those transformations, Ω, which enter into the F–P condition (4). Then we can write

$$\langle TF(g, \Gamma)\rangle_{\Phi_1} = \int (dg d\Gamma) F(g, \Gamma) \exp\left((i/\hbar)(\mathscr{S}_P(g, \Gamma) + \mathscr{S}_{\Phi_1}(g, \Gamma))\right)$$

$$\int (d\Omega) \exp\left((i/\hbar)\mathscr{S}_{\Phi_2}(g^{\Omega^{-1}}, \Gamma^{\Omega^{-1}})\right)$$

$$= \int (d\Omega dg d\Gamma) F(g^\Omega, \Gamma^\Omega) \exp\left((i/\hbar)(\mathscr{S}_P(g, \Gamma) + \mathscr{S}_{\Phi_1}(g^\Omega, \Gamma^\Omega) + \mathscr{S}_{\Phi_2}(g, \Gamma))\right)$$

after changing orders of integration and making a change of integration variables. This formula can be read as the transformation rule,

$$\langle TF(g, \Gamma)\rangle_{\Phi_1} = \int (d\Omega)\langle TF(g^\Omega, \Gamma^\Omega) \exp\left((i/\hbar)(\mathscr{S}_{\Phi_1}(g^\Omega, \Gamma^\Omega))\right)\rangle_{\Phi_2}, \quad (6)$$

which clearly includes the rule for transforming Green's functions. The F–P condition played an essential role in its derivation.

While the transformation rule (6) is complicated for Green's functions in general, it takes a very simple form for those functionals which are invariant under general coordinate transformations. Thus, if

$$F(g^\Omega, \Gamma^\Omega) = F(g, \Gamma),$$

then (6) and (4) imply

$$\langle TF(g, \Gamma)\rangle_{\Phi_1} = \langle TF(g, \Gamma)\rangle_{\Phi_2}. \quad (7)$$

That is, the quantum amplitudes which correspond to invariant classical functionals are themselves *gauge independent*. In particular, the on-shell S-matrix belongs to this category.

There is a good deal of latitude in the choosing of the supplementary action.[12] The F–P condition can always be satisfied by taking an arbitrary (but non-covariant) functional $\mathscr{S}_\Phi{}^{(0)}$ and adjoining to it a suitable covariant term $H(g, \Gamma)$. Thus, if one writes

$$\mathscr{S}_\Phi = \mathscr{S}_\Phi{}^{(0)} + H, \tag{8}$$

where H is given by the integral

$$\exp\left(-(i/\hbar)H(g, \Gamma)\right) = \int(d\Omega) \exp\left((i/\hbar)\mathscr{S}_\Phi{}^{(0)}(g^\Omega, \Gamma^\Omega)\right), \tag{9}$$

then the F–P condition (4) is clearly satisfied. For many choices of $\mathscr{S}^{\Phi(0)}$ the integral (9) can be evaluated explicitly.

4. PERTURBATION EXPANSIONS

Although integrals such as (5) and (9) have not been defined with any precision it is still possible to gain some insight into their general structure by perturbation methods. A formal application of the standard perturbation technique yields for these integrals series of terms which can be put into correspondence with Feynman graphs in the usual way. Owing to the formal nature of this procedure there will of course result many ambiguous or singular expressions in the perturbation series.

The Palatini action density is given explicitly by

$$\mathscr{L}_P = (1/\kappa^2)\sqrt{-g}\, g^{\mu\nu}[\Gamma^\lambda{}_{\lambda\nu,\,\mu} - \Gamma^\lambda{}_{\mu\nu,\,\lambda} + \Gamma^\rho{}_{\lambda\nu}\Gamma^\lambda{}_{\mu\rho} - \Gamma^\rho{}_{\mu\nu}\Gamma^\lambda{}_{\rho\lambda}], \tag{10}$$

where $\kappa^2/8\pi = G$, the Newtonian constant. To obtain expansions in powers of κ, one substitutes for $g_{\mu\nu}$ and $\Gamma^\lambda{}_{\mu\nu}$ the variables $\phi_{\mu\nu}$ and $\phi^\lambda{}_{\mu\nu}$ defined by

$$g_{\mu\nu} = \eta_{\mu\nu} + \kappa\phi_{\mu\nu} \quad \text{and} \quad \Gamma^\lambda{}_{\mu\nu} = \kappa\phi^\lambda{}_{\mu\nu}, \tag{11}$$

where $\eta_{\mu\nu}$ denotes the Minkowskian tensor. (This is only one of many possible parametrizations of the gravitational fields. In general one can take any functions $g_{\mu\nu}(\phi)$, $\Gamma^\lambda{}_{\mu\nu}(\phi)$ which reduce in the limit $\phi = 0$ to some specified solution of the Einstein equations.) The total action, $\mathscr{S}(g, \Gamma)$, when expressed in terms of the new fields, ϕ, takes the form

$$\mathscr{S} = \mathscr{S}_0(\phi) + \mathscr{S}_{int}(\phi), \tag{12}$$

where $\mathscr{S}_0(\phi)$, the 'bare' action, is a bilinear functional of the ϕs and is independent of κ. This term, whose exact form depends upon the choice of supplementary action $\mathscr{S}_\Phi$, determines the bare graviton propagator. The remaining part, $\mathscr{S}_{int}(\phi)$, takes the form of a power series in κ,

$$\mathscr{S}_{int}(\phi) = \kappa\mathscr{S}_{(1)}(\phi) + \kappa^2\mathscr{S}_{(2)}(\phi) + \dots \tag{13}$$

The functionals $\mathscr{S}_{(n)}(\phi)$ determine the vertex structure. Their forms also depend upon the choice of supplementary action. A typical perturbation series is obtained from the path-integral representation,

$$\langle T\phi(x_1)\phi(x_2)\ldots\rangle = \int(d\phi)\phi(x_1)\phi(x_2)\ldots\exp\left((i/\hbar)(\mathscr{S}_0(\phi)+\mathscr{S}_{int}(\phi))\right)$$

by expanding the factor $\exp(i/\hbar)\mathscr{S}_{int}(\phi)$ in powers of κ. The resulting terms have the form of Gaussian integrals (since $\mathscr{S}_0(\phi)$ is bilinear) and can therefore be evaluated. One obtains in this way the usual Feynman rules which are familiar from the operator formulations of perturbation theory.[13]

The new and indeed surprising feature of the perturbation series in quantized general relativity lies in the structure of the interaction terms $\mathscr{S}_{(n)}(\phi)$ which – for most choices of gauge – are non-local. That is, they yield vertices in which the graviton lines do not all emerge from a single space–time point but rather from distinct points which are themselves connected by lines which can be thought of as corresponding to the propagation of virtual 'fictitious' particles.

The fictitious particle contributions are contained in the term $H(\phi)$ whose structure is determined by the integral (9). Thus, even if one chooses $\mathscr{S}_\Phi^{(0)}$ to be local (and independent of $\hbar$, i.e. purely classical) one finds by imposing the F–P condition that the supplementary action must contain a part which is non-local and $\hbar$-dependent. Since (9) has the form of a path-integral representation of a quantized amplitude its structure can also be analysed graphically. To do this, the integration variables $\Omega^\mu(x)$ should be parametrized in terms of some suitable field $B^\mu(x)$ and the (local) action $\mathscr{S}_\Phi^{(0)}(g^\Omega, \Gamma^\Omega)$ is then expressed in powers of B,

$$\mathscr{S}_\Phi^{(0)} = \mathscr{A}_{(0)}(B) + \mathscr{A}_{int}(B, \phi) \tag{14}$$

where $\mathscr{A}_0(B)$ is bilinear in B and independent of ϕ. By expanding the integrand of (9) in powers of $\mathscr{A}_{int}$ one obtains a series of Gaussian integrals which can, in principle, be evaluated. The terms of this series can be put into correspondence with Feynman graphs whose lines represent the bare propagator (determined by $\mathscr{A}_0(B)$) of the fictitious field $B^\mu(x)$ and whose vertices are determined by $\mathscr{A}_{int}(B, \phi)$. The B-lines are all internal. It can be shown that the functional $(-i/\hbar)H(\phi)$ is represented by the sum of all such graphs which are B-connected, i.e. which cannot be separated into two pieces without cutting a B-line. A diagram with L closed loops will represent a contribution proportional to $\hbar^L$ in $H(\phi)$. This is the origin of the non-classical terms in the total action $\mathscr{S}(\phi)$.

Finally, by treating the internal ϕ- and B-lines on the same footing, one can set up Feynman rules in which all the vertices are local. To take account of the fact that the total action (12) involves $+H(\phi)$, while the rules outlined in the previous paragraph give $-H(\phi)$, it is necessary to multiply the contribution of any graph by the factor $(-)^N$, where N denotes the number of B-connected pieces in it.

From this brief outline one can see that the perturbative structure of amplitudes in quantized general relativity theory is very much complicated by the presence of the fictitious quanta – whose role is to make gauge-independent those amplitudes (such as the on-shell S-matrix) which should be. On the other hand, by suitable choice of gauge one can effect considerable simplifications. We mention two examples.

Both of these examples take $\mathscr{S}_\Phi{}^{(0)}$ of the form,

$$\mathscr{S}_\Phi{}^{(0)} = \int d_4 x C_\mu \Phi^\mu(g, \Gamma), \tag{15}$$

where C_μ denotes a supplementary (Lagrange multiplier) field which must be included, along with $g_{\mu\nu}$ and $\Gamma^\lambda{}_{\mu\nu}$, among the integration variables in (1). The F–P condition takes the form

$$\exp\left(-(i/\hbar)H(g, \Gamma)\right) = \int(d\Omega dC)\exp\left((i/\hbar)\int dx C_\mu \Phi^\mu(g^\Omega, \Gamma^\Omega)\right)$$
$$= \int(d\Omega)\delta(\Phi^\mu(g^\Omega, \Gamma^\Omega)). \tag{16}$$

This integral can be evaluated on the subspace $\Phi^\mu = 0$ since only infinitesimal transformations, Ω, will contribute. For $\Omega^\mu = x^\mu + \epsilon^\mu$ we have, to first order in ϵ,

$$\Phi^\mu(g^\Omega, \Gamma^\Omega) = \Phi^\mu(g, \Gamma) - \epsilon^\nu \partial_\nu \Phi^\mu(g, \Gamma) + Q^\mu{}_\nu \epsilon^\nu,$$

where Q denotes a differential operator whose coefficients depend on g and Γ. The integral (16) reduces to

$$\exp\left(-(i/\hbar)H(g, \Gamma)\right) = \int(d\epsilon)\delta(Q^\mu{}_\nu \epsilon^\nu)$$
$$= |\mathrm{Det}\, Q^\mu{}_\nu|^{-1}, \tag{17}$$

which is a functional determinant. Equivalently,

$$H(g, \Gamma) = -i\hbar Tr \ln Q^\mu{}_\nu + \text{constant}, \tag{18}$$

where the constant is conveniently fixed to give $H = 0$ in flat space–time. The examples are:

1. DE DONDER–LANDAU GAUGE

$$\Phi^\mu = g^{\lambda\rho}\, \Gamma^\mu{}_{\lambda\rho} \tag{19}$$

v

$$Q^\mu{}_\nu = -\delta^\mu{}_\nu g^{\lambda\rho} \partial_\lambda \partial_\rho$$

$$= -\delta^\mu{}_\nu \left(1 + \kappa\phi^{\lambda\rho}\frac{\partial_\lambda\partial_\rho}{\partial^2}\right)\partial^2, \tag{20}$$

where the last expression refers to the parametrization $g^{\mu\nu} = \eta^{\mu\nu} + \kappa\phi^{\mu\nu}$ and $\partial^2 = \eta^{\mu\nu}\partial_\mu\partial_\nu$. The resulting expression for H,

$$H = -i\hbar \sum_{n=1}^\infty \frac{\kappa^n}{n} \int dx_1 \ldots dx_n\, \phi^{\mu_1\nu_1}(x_1)\, \partial_{\mu_1}\, \partial_{\nu_1}\, D(x_1 - x_2)$$

$$\times \phi^{\mu_2\nu_2}(x_2)\, \partial_{\mu_2}\, \partial_{\nu_2}\, D(x_2 - x_3) \ldots \phi^{\mu_n\nu_n}\, \partial_{\mu_n}\, \partial_{\nu_n}\, D(x_n - x_1), \tag{21}$$

corresponds to the propagation of a fictitious vector particle around simple closed loops. For each $n \geqslant 1$ there is essentially only one connected graph, a fictitious particle loop with n vertices at each of which one graviton is emitted.

2. GAUSSIAN GAUGE

$$\Phi^\mu = g^{0\mu} - \delta^\mu{}_0 \tag{22}$$

$$Q^\mu{}_\nu = (\delta^0{}_\nu g^{\mu\alpha} + \delta^\mu{}_\nu g^{0\alpha})\, \partial_\alpha. \tag{23}$$

It can be shown in this case that Det $Q = 1$ on the subspace $\Phi = 0$. In other words, H vanishes and there are no fictitious particles in this gauge.

5. UNITARITY AND INTER-RELATIONSHIPS OF THE CANONICAL AND PATH INTEGRAL QUANTIZATION METHODS

Although the fictitious particles were deduced above in the context of gauge-independence, they also play a role in the unitarity of the S-matrix. It was from this viewpoint that Feynman originally came upon them.[14] A possible way to 'demonstrate' unitarity would be to reduce the path integrals (1) to canonical form, showing thereby the existence of an effective Hermitian Hamiltonian. Unfortunately we have not been able to carry out this programme since some of the integrals involved are not Gaussian. (This difficulty corresponds to the occurrence of the non-linear constraint equations which prevent Arnowitt, Deser and Misner from obtaining an explicit form for the classical Hamiltonian.) On the other hand, we can carry out the reduction to the canonical form of Dirac which involves six pairs of canonical variables subject to some constraints. However, since the path integrals can shed no light on the ordering problems when treated in the formal manner we are using, such a reduction gives no information about the consistency of the constraints.

To illustrate the reduction to canonical form in a simpler theory, we consider now the Yang–Mills field, which bears some resemblance to gravity in that it possesses a non-Abelian gauge symmetry of the second kind. Here the canonical reduction can be carried through explicitly. We shall sketch the procedure for the Yang–Mills field and then return, briefly, to the problem of gravity.

The covariant part of the Yang–Mills action density is given by

$$\mathscr{L}_{YM} = -\tfrac{1}{2}F^k{}_{\mu\nu}(\partial_\mu A^k{}_\nu - \partial_\nu A^k{}_\mu + gf^{klm}A^l{}_\mu A^m{}_\nu) + \tfrac{1}{4}F^k{}_{\mu\nu}F^k{}_{\mu\nu}, \quad (24)$$

where k, ℓ, m denote internal symmetry indices. Under an infinitesimal Yang–Mills transformation the fields transform according to

$$\delta A^k{}_\mu = f^{klm}A^l{}_\mu \epsilon^m + \frac{1}{g}\partial_\mu \epsilon^k$$

$$\delta F^k{}_{\mu\nu} = f^{klm}F^l{}_{\mu\nu}\epsilon^m. \quad (25)$$

The supplementary part of the action (corresponding to the classical radiation gauge condition $\partial_a A^k{}_a = 0$) takes the form

$$\exp((i/\hbar)\mathscr{S}_\phi(A)) = \delta(\partial_a A^k{}_a)\exp((i/\hbar)H(A)), \quad (26)$$

where $H(A)$ denotes a Yang–Mills invariant functional which is determined by the F–P condition, i.e.

$$\exp(-(i/\hbar)H(A)) = \int(d\Omega)\delta(\partial_a A^\Omega{}_a)$$
$$= \text{constant} \times \exp(Tr \ln \mathscr{D}(x, x'|A)), \quad (27)$$

where the constant is chosen to make $H(0) = 0$. One can obtain the Green's function $\mathscr{D}$ by solving the equation[15]

$$\nabla^2 \mathscr{D}^{kn} + gf^{klm}A^l{}_a\partial_a \mathscr{D}^{mn} = \delta^{kn}\delta(\mathbf{x}-\mathbf{x}').$$

The solution can be obtained in powers of g. The term of order n corresponds to a graph in which a fictitious scalar, static, particle forms a closed loop with n vertices at each of which a Yang–Mills field is attached.

In order to see the relevance to unitarity of the fictitious particle contributions in $H(A)$ it is necessary to reduce the basic path integral to canonical form by integrating over all redundant field components. Since the canonical variables are the transverse parts of $A^k{}_a$ and $E^k{}_a = F^k{}_{a0}$, the redundant components are $A^k{}_0$, $F^k{}_{ab}$ and the longitudinal component of $F^k{}_{a0}$. All these redundant integrals are Gaussian and can be done explicitly. The resulting path integral has the form,

$$\int(dA_a)(dE_a)\delta(\partial_a A_a)\delta(\partial_a E_a)\exp((i/\hbar)\int dx(E_a\dot{A}_a - H)), \quad (28)$$

where the Hamiltonian density, H, is given by

$$H = \tfrac{1}{2}(E^k{}_a)^2 + \tfrac{1}{4}(\partial_a A^k{}_b - \partial_b A^k{}_a + g f^{klm} A^l{}_a A^m{}_b)^2 +$$

$$+ \frac{g^2}{2} A^k{}_a E^l{}_a f^{klm} \mathscr{D}^{mn} \nabla^2 \mathscr{D}^{nrfrst} A^s{}_b E^t{}_b. \tag{29}$$

The last term here is non-local but classical (i.e. independent of $\hbar$). The non-classical factor $H(A)$ has disappeared in the course of integrating over the redundant components. This means that the Faddeev–Popov prescriptions are equivalent, in the radiation gauge, to the traditional method of canonical quantization. Hence, at least to that extent, they must yield a unitary S-matrix.

For the gravitational problem a suitable gauge is the one mentioned in section 3, which corresponds to classical Gaussian coordinates. In this gauge the supplementary action is given by

$$\exp((i/\hbar)\mathscr{S}_\Phi) = \delta(g^{0\mu} - \delta^\mu{}_0). \tag{30}$$

The corresponding path integral representation is given or, rather, symbolized by

$$\langle TF(g, \Gamma) \rangle = \int (dg \, d\Gamma) \delta(g^{0\mu} - \delta^\mu{}_0) F(g, \Gamma)$$

$$\times \exp((i/\hbar)\mathscr{S}_P(g, \Gamma) - 3 \int dx \delta(x, x) \ln \sqrt{-g}), \tag{31}$$

where $\mathscr{S}_P$ denotes the Palatini action. The term involving the singular quantity $\delta(x, x) = \delta(x - x)$ has been introduced in order that the integral shall reduce to canonical form. Integration over the components $g_{0\mu}$, $\Gamma^l{}_{mn}$ and $\Gamma^\mu{}_{0\nu}$ with $\Gamma^0{}_{mn}$ replaced by π^{ab} according to the definition

$$\Gamma^0{}_{mn} = \frac{1}{\sqrt{-g}} (\tfrac{1}{2} g_{mn} g_{ab} - g_{ma} g_{nb}) \pi^{ab}, \tag{32}$$

leads to the expression

$$\int (d_6 g \, d_6 \pi) \exp((i/\hbar) \int dx (\pi^{mn} \dot{g}_{mn} - H)), \tag{33}$$

where the Hamiltonian density, H, is given by

$$H = \tfrac{1}{2} K^{-1}(g_{im} g_{jn} + g_{in} g_{jm} - g_{ij} g_{mn}) \pi^{ij} \pi^{mn} - K^{(3)}R, \tag{34}$$

where $K = \sqrt{(-\det g_{mn})}$ and $^{(3)}R$ denotes the three-dimensional curvature scalar. The steps which led to (33) were purely formal and took no account of the ordering question. Moreover the constraint equations of Dirac are not manifestly satisfied. Presumably the field equations which are derivable from the Hamiltonian (34) are sufficient to ensure that the constraints will be preserved in time if they are imposed initially. However, these initial conditions must be reflected in the asymptotic properties of the integration domain in (33) about which we know nothing.

Another aspect of the formal nature of these considerations is the appearance of the singular $\delta(x, x)$ in (31). No doubt the expression

$$\int(dg) \exp\left(-3 \int dx \delta(x, x) \ln\sqrt{-g}\right)$$

should be interpreted as a limit of the multiple integral

$$\prod_x \int (-g(x))^{-3/2} \prod_{\mu \leqslant \nu} dg_{\mu\nu}(x)$$

taken over a net of points x, with the net becoming dense in the limit.[16]

To summarize, the use of path-integral representations appears to offer a flexible and potentially useful approach to the treatment of the quantized gravitational field. The ease with which redundant variables can be introduced or integrated out gives a comprehensive view of equivalent formulations: canonical quantization with or without constraints, and covariant quantization in various gauges. In the operator formulations of field theory it is very difficult to trace these connections. Thus, many aspects of the quantization problem are formally comprehended by the path-integral approach and there is present the possibility that the deeper consistency problems can be resolved once a more precise definition of the integral is obtained.

NOTES AND REFERENCES

1. A discussion of these developments is contained in the articles by B. S. DeWitt, *Phys. Rev.* **160**, 1113 (1967); **162**, 1195, 1239 (1967). See also the survey by T. W. B. Kibble in *High-Energy Physics and Elementary Particles*, p. 885, Trieste Seminar (IAEA, Vienna: 1965).
2. P. A. M. Dirac, *Contemporary Physics*, vol. 1, p. 539, Trieste Symposium (IAEA, Vienna: 1968).
3. R. P. Feynman, *Rev. Mod. Phys.* **20**, 367 (1948).
4. P. A. M. Dirac, *Can. J. Math.* **2**, 129 (1950); *Proc. Roy. Soc. (London)* A246, 326, 333 (1958); *Phys. Rev.* **114**, 924 (1959).
5. R. Arnowitt, S. Deser and C. W. Misner, *Phys. Rev.* **113**, 745 (1959); **116**, 1322 (1959); **117**, 1595 (1960).
6. J. Schwinger, *Phys. Rev.* **132**, 1317 (1963).
7. S. N. Gupta, *Proc. Phys. Soc. (London)* A65, 608 (1952).
8. B. S. DeWitt, *Phys. Rev.* **162**, 1195, 1239 (1967).
9. S. Mandelstam, *Phys. Rev.* **175**, 1604 (1968).
10. R. P. Feynman, ref. 3. P. T. Matthews and Abdus Salam, *Nuovo Cimento* **2**, 120 (1955); R. P. Feynman and A. R. Hibbs, *Quantum Mechanics and Path-Integrals* (McGraw-Hill, New York: 1965).
11. Path-integral methods have been applied to the gravitational problem by B. Laurent, *Nuovo Cimento* **4**, 1445 (1956); *Arkiv Fysik* **16**, 279 (1959); C. W. Misner, *Rev. Mod. Phys.* **29**, 497 (1957); J. R. Klauder, *Nuovo*

Cimento **19**, 1059 (1961); B. S. DeWitt, *J. Math. Phys.* **3**, 1073 (1962) and ref. 7; R. P. Feynman, *Acta Phys. Polon.* **24**, 697 (1963); H. Leutwyler, *Phys. Rev.* **134**, B1155 (1964); L. D. Faddeev and V. N. Popov, Kiev preprint ITF 67-36 (1967) (in Russian); *Phys. Letters* **25B**, 29 (1967).

12. It is often convenient to introduce supplementary fields into the definition of $\mathscr{L}_\Phi$. These fields act as Lagrange multipliers, cf. equation (15).

13. P. T. Matthews and Abdus Salam, ref. 10.

14. R. P. Feynman, ref. 11.

15. E. S. Fradkin and I. V. Tyutin, *Phys. Rev.* D2, 2841 (1970). This article contains an alternative approach to the quantization of non-Abelian gauge theories.

16. The measure question is a vexed one. Various proposals have been put forward in the articles listed in ref. 11. Our own suggestion is highly tentative.

Index

INDEX

265

w